FOSSILS FOR AMATEURS

FOSSILS FOR AMATEURS

A Guide to Collecting
and Preparing Invertebrate Fossils

RUSSELL P. MacFALL
& JAY WOLLIN

We are the Ancients of the Earth,
And in the morning of the times.
Tennyson

Van Nostrand Reinhold Company
NEW YORK CINCINNATI TORONTO LONDON MELBOURNE

Dedicated to all the fossils yet unfound

Van Nostrand Reinhold Company Regional Offices:
New York   Cincinnati   Chicago   Millbrae   Dallas
Van Nostrand Reinhold Company International Offices:
London   Toronto   Melbourne

Copyright © 1972 by Litton Educational Publishing, Inc.
Library of Congress Catalog Card Number 74-149259

Published by Van Nostrand Reinhold Company
450 West 33rd Street, New York, N.Y. 10001
Published simultaneously in Canada by
Van Nostrand Reinhold Limited
16 15 14 13 12 11 10 9 8 7 6 5 4 3 2 1

FOREWORD

Some years ago, I clipped a sentence from an abstract in the program of the annual meeting of the Geological Society of America, and pasted it to the bookshelf by my desk. I no longer remember who was the author of this quotation, and I don't know whether his tongue was in his cheek when he wrote: "Geologists are the most literate of all technical writers. Skilled in a descriptive science, strong on grammar, they are beautifully, even romantically fluent."

Perhaps there is an affinity between the earth and writers—and between writers and the earth. Perhaps the thrills and pleasures of digging up clues to the history of our planet—tangible bits of the planet itself—require to be shared.

This sharing can be seen on many levels. Specialists, popularizers, tyros, teachers, hobbyists: all have something to say, and all to more than one audience. Russell MacFall, for many years an editor of the *Chicago Tribune,* is a hobby collector of both fossils and minerals. His amateur standing with respect to the latter category was shattered two decades ago with the publication of his *Gem Hunter's Guide.* His amateur position in paleontology is shattered by the volume now before us. Jay Wollin, briefly a landscape designer, holds degrees in horticulture and paleobotany and currently teaches earth science at Oakton Community College in Morton Grove, Illinois.

Both authors remain authentic members of the far-flung community of amateurs of paleontology; for both remain keen collectors and happy devotees of the thrill of uncovering and preserving fossils. It is to their fellow collectors that they address this book. In presenting it, the publisher has chosen well, for one could look long and far to find authors so well established in the affections of their fellows. Both have come up through the ranks of local collecting societies to the presidency of the Midwest Federation of Mineralogical and Geological Societies.

Eugene S. Richardson Jr.
Curator, Fossil Invertebrates,
Field Museum of Natural History,
Chicago, Illinois

ACKNOWLEDGMENTS

The authors wish to acknowledge a debt of gratitude to Dr. Richardson and to the Field Museum of Natural History which they can never hope to repay. Dr. Richardson read critical parts of the manuscript, masterfully extricating the authors from pitfalls of fact and phraseology into which they had blundered. The museum staff patiently gave access to fossil collections and preparation rooms and gave freely of their own ideas.

The book also owes much of its value to the close cooperation of Mrs. Betty Crawford of Mansfield, Ohio, who supplied photographs and made pencil drawings and detailed state maps for it. Dr. Dwayne Stone of Marietta College, Dr. Kenneth E. Caster of the University of Cincinnati, and Dr. Frank M. Carpenter of Harvard University generously lent photographs; and a number of state geological surveys supplied useful material.

Hands appearing in some photographs are visual evidence of the cooperation of Alice Wollin, Philip Olsberg, and Orville Gilpin. The manuscript was neatly retyped several times by Mrs. Diane Bowden, Miss Nadine Abicht, and Miss Beth Burgess. Photographs, unless otherwise credited, were taken by Jay Wollin and printed by Michael Gibson and Tom McCleary.

Finally, the authors wish to thank the several editors of our publisher who have guided the manuscript to completion, and especially Patricia Horgan of New York, who has the supreme wisdom to know how and when to ask the right questions.

CONTENTS

TREASURES IN THE DUST

Some fishermen have all the luck—like Francis Tully. Tully's favorite fishing spot is in the little lakes that punctuate the strip-coal-mining region not many miles south of his home in Lockport, Illinois, near Chicago. When the fish weren't biting, Tully passed the time by breaking open some of the rusty brown ironstone concretions that littered the huge dumps of waste rock.

Occasionally he would find a fossil fern frond inside a concretion, and one day in the early 1960s he found his first fossil shrimp and, later, a worm. Tully, a pipefitter at a Lockport oil refinery, was curious enough to seek further information about his discoveries, which brought him and his fossils to the attention of paleontologists at the Field Museum of Natural History in Chicago.

Then came a memorable day. Fishing was poor but the red-brown nodules were plentiful. Tully cracked open a few, hoping to find another fern or, as he once had, a dragonfly entombed in the rock. That discovery had cost him a battered thumb from a careless blow of his hammer. But this time what he found was not fern or insect.

What was this strange outline in the rock that he held in his hand? The creature, if it was a creature, had an elongated, cigar-shaped body divided into segments. At one end was a spadelike tail and on the other a slender snout ending in a toothed claw. Behind the snout a bar terminated by two small lumps crossed the body. The strange creature was about five inches long.

Tully was puzzled. So were the experts when he took his discovery to the Field Museum. They couldn't find a place for it in the fossil picture—the jigsaw puzzle of life scrambled in the rocks. Like all scientists, they were primarily concerned with recognizing and then classifying the object into a coherent body of knowledge. Customarily they do this by comparing an undescribed fossil with known, already classified, forms until relationships can be established. Then the fossil takes its place in the scheme of things.

Dr. Eugene S. Richardson, Jr., the museum's specialist in invertebrate fossils, was intrigued by what a visitor called the "impish, benevolent, almost schmoolike expression on its cuddly frame," but he was also baffled by the fossil's enigmatic appearance. Even a stranger needs a name, so he named it *Tullimonstrum gregarium* (common Tully monster), using the discoverer's name in Latinized form for the genus and a common Latin adjective for the trivial name. But even with a name, Tully's monster lacked a place; it did not even appear to belong to any known phylum, which is one of the most general of all groups of classification. It was an orphan in the family of things, living and dead.

Since that day, Tully has been back to the site of his great discovery hundreds of times and has collected 3,000 animal fossils there, including 150 Tully monsters and nearly as many sea cucumbers, as well as several hundred worms of several species. He keeps them carefully wrapped in paper towels in cabinet drawers.

Scores of other amateurs have swarmed over the strip-mine spoil piles. They, too, have unearthed Tully monsters and added them and other strange fossils to their collections. Many of these collectors have shared their specimens with Dr. Richardson. From the evidence that such cooperation has made available, he deduced that the monster was an animal, a marine animal that preyed on other life in an ancient sea. It may have fed on the strange creatures that are dubbed "blobs" because of their variable shapes, somewhat like wilted lettuce leaves. These creatures are found in the strip-mine concretions and are as little understood as the monster itself.

When the national paleontological convention was held at the Field Museum in the fall of 1969, hundreds of professional paleontologists from all over the world saw a special exhibition of cases full of the specimens that amateurs had made available to the museum.

Most of these scholars had heard of the water-filled pits and sterile gray clay hills south of Chicago left by the huge dragline buckets as they stripped the farmland to bare the coal seams. This area had long been world-famous for the variety of well-preserved Coal Age fossils found there, especially the fossil ferns and other plants. As long ago as the mid-nineteenth century, eager amateurs had waded in shallow Mazon Creek, sixty miles southwest of Chicago, to gather such ironstone concretions.

Systematic studies were based on the specimens they collected, and major museums of Europe and the United States placed these fossils in their cabinets. But the fossils really became abundant—by the thousands and hundreds of thousands—when the strip miners brought their giant draglines to the cornfields after World War I.

The coal beds were formed in the Middle Pennsylvanian period about 300 million years ago. At that time, forests grew on a featureless coastal plain that lay between a sea to the southwest and a hilly upland far to the northeast. Giant tree ferns and horsetails, ancestors of the puny present-day representatives of these families, and other vegetation flourished in the swampy plain between sea and hills. Trees fell, plant life died, and the advancing sea buried the debris under layers of mud. Instead of rotting away, as it would have done if it had been exposed to the air, the vegetable mass accumulated in water and then was compressed by the blanket of mud and sand above it while it slowly turned into fossil fuel.

Tully's monster and its exotic playmates apparently lived in the swamps and shallow offshore waters, with sand bars and mud flats. It was a setting rich in animal life. Pectens and snails, sea cucumbers, horseshoe crabs, annelid worms, and even rare amphibians lived here. Insects cruised overhead, and the little monster paddled its way through the murky, milk-warm waters along with sharks, primitive fish, shrimp, and squid. Like the trees, these animals died and their bodies sank to the muddy bottom. Products of decay caused iron to precipitate from the sea water and to harden the mud around their bodies in concentric layers. Seas rose and fell, the coal formed, the mud hardened into shale, glaciers planed the land and left behind a fertile prairie. The concretions waited 300 million years until someone finally tapped the rusty coffins and exposed their flattened ghosts to the open air again.

Besides the monster, the strip-mine dumps yielded other prized fossils— jellyfish. These are rare in any rocks because the soft, watery tissues of jellyfish leave a fossil record only under the most favorable conditions. They are exceptional fossils, and paleontologists prize them because paleontologists, like other people, treasure the improbable. Amateurs again made the two big discoveries. Jim Turnbull of Libertyville, a suburb north of Chicago, brought to the Field Museum the fossil of a jellyfish four

Fossilized squid from a coal mine concretion near Braidwood, Illinois. It was named *Jeletzkya douglassae* for the young amateur who found it.

inches across. Like Tully, he was honored in the name given it—*Anthracomedusa turnbulli*, which in plain English means "Coal Age jellyfish of Turnbull." Dr. Richardson also discovered tiny jellyfish among fossils in the collection of the Ted Pieckos of Chicago. These, too, had come from the strip mines. He named them *Octomedusa pieckorum*, or "eight-sided jellyfish of the Pieckos." A museum visitor, one of the cooperating amateur collectors, viewed pictures of the jellyfish and came back a week later with 400 specimens!

The Pieckos later topped their jellyfish discovery by bringing to the museum the first fossil lamprey ever found, one of seven they had col-

Helen and Ted Piecko of Chicago, discoverers of several new fossils and owners of one of the largest collections of fossils from the strip mines southwest of Chicago. (Photo courtesy Mr. and Mrs. Piecko)

lected at the same strip mines that had yielded the jellyfish. It, too, was named in their honor.

Tully, Turnbull, and the Pieckos are only examples of the hundreds of amateur collectors in the Chicago area who have "pursued their hobby and the advancement of science at the same time," as Dr. Richardson says. In this way amateurs and professionals help each other to read the story of the past written in the rocks that lie almost in the shadow of a great city's skyscrapers.

Finding marine monsters in a coal mine and jellyfish on a barren hillside a thousand miles from the sea is strange enough, but no stranger than gathering a bouquet of 325-million-year-old lilies in a railroad stone quarry. Crop after crop, too! Burnice H. Beane of Le Grand, Iowa, a singularly frail-looking man for such a career, spent his lifetime cultivating his lily garden. He was born in 1879, five years after a blast in the quarry at Le Grand laid bare a bed of "stone flowers" on the dislodged limestone slabs. These were fossils of crinoids, marine animals related to starfish but bearing a deceptive likeness to daisies or lilies.

Beane spent many days of his boyhood in the quarry, asking questions of the workmen and of collectors, and gathering crinoids himself until

B. H. Beane with a slab of crinoid fossils from a quarry at Le Grand, Iowa. (Photo Des Moines *Register & Tribune*)

In honor of the Freys, the Harvard scientists named the amber-entombed insects *Sphecomyrma freyi*, or "wasp-ant of Frey."

These examples mean something more than the discovery of fossil species, important though that is. Tully might have tossed his strange fossil back into the pond, or the Freys might have placed their bit of amber on the shelf. But they didn't, because they possessed a persistent trait of the true collector—stubborn curiosity. They were curious enough to want to know more, to learn all they could about their specimens and to deepen their understanding of the science of paleontology and the craft of collecting.

Collectors learn in many ways. They learn from other amateurs, from professional paleontologists, and from books, which preserve the accumulated experience and knowledge mankind has painfully won. When the young Charles Darwin went aboard the *Beagle* in 1831 he carried with him Sir Charles Lyell's newly published *Principles of Geology*. While he sailed around the world, Darwin read, and what he learned from Lyell's

Sphecomyrma freyi, in amber, the primitive ant that appears to be a link between ants and wasps. (Photo courtesy Dr. F. M. Carpenter)

book opened his eyes to new meanings in the fossils he discovered in South America and the Galapagos Islands. From understanding—and genius—grew the theory of evolution.

The book you are now reading presents the essential background and the practical techniques that will enable the amateur collector to get more pleasure and intellectual satisfaction from his activities. As a glance at the table of contents shows, it describes what a fossil is and how to distinguish it from other rocks. It speaks briefly about the vast extent of geologic time and the patient workings of geologic processes, and it maps, state by state, the rocks that are likely to contain fossils, with particular emphasis on areas known to be productive.

From such basic information, the book turns to preparations for a collecting trip, interpretation of maps, and a summary of state and federal laws regulating the collecting of fossils. Once the fossil is collected it must be prepared for study or display; so the book describes the latest techniques for freeing fossils from the matrix and protecting them on the long trip home. It also describes in detail how to clean, prepare, preserve, exhibit, and photograph specimens.

The last chapter outlines the procession of life from the most primitive plants and animals to those invertebrates that prepared the way for the development of the mammals. It mentions the major fossils the amateur will be likely to collect and is designed to serve him as a quick reference source for identifying and classifying his treasures.

The scope of the book has purposely been limited to invertebrate fossils for two reasons: first, because these are generally the fossils that amateurs collect, and second, because in collecting, preparing, and exhibiting vertebrate fossils—especially the bones and teeth of large animals —problems are presented that can be solved only with the facilities of a museum or university. The amateur who collects vertebrate fossils may destroy material that is of major scientific value and may also subject himself to penalties for violating the law.

Dr. J. R. Macdonald, curator of vertebrate paleontology at the Los Angeles County Museum, offers sound counsel on this point. After citing instances of destruction of vertebrate fossils by thoughtless or unskilled collectors, he writes: "One would almost get the feeling that I am against rockhounds, pebble pups and the rest of that large group of hobbyists. This is not true; rather than fight them I'd much rather see them join us. It is from these outdoor enthusiasts that we get much of our scientifically valuable material and leads on sites which help us to expand our knowledge of life in the past."

Dr. Macdonald mentions several important discoveries made through this kind of cooperation in California. Nor was the gain all on the side of the professionals, for the amateurs enjoyed the experience of helping to excavate a major fossil, of working on it in the laboratory, and even, in some instances, of having fossils named for them.

"To collect most fossil vertebrates properly," Macdonald writes, "a great deal of time, training and skill is required. Once collected, it is not a scientific specimen but just a curiosity unless the specimen is accompanied by complete geographical and geological data. . . . Why can't the hobbyist join us in our search for knowledge and in the reconstruction of the history of earth's past?"

There is plenty of room for the amateur to have his fun and the scientist his knowledge. For, as the eminent Harvard paleontologist George Gaylord Simpson once wrote: "Fossil hunting is far the most fascinating of all sports. . . . The hunter never knows what his bag may be, perhaps nothing, perhaps a creature never before seen by human eyes. Over the next hill may lie a great discovery. It requires knowledge, skill, and some degree of hardihood. And its results are so much more important, more worthwhile, and more enduring than those of any other sport. The fossil hunter does not kill; he resurrects. And the result of his sport is to add to the sum of human pleasure and to the treasures of human knowledge."

THE NATURE OF FOSSILS

Simply defined, fossils are the remains or traces of organisms that lived prior to historic times. They most commonly preserve the shape or impression of the organism itself in rock, but they may be actual bone, or flesh preserved by freezing, or trails and other marks made by ancient animals. Today it is generally accepted that life has existed on our earth for more than two billion years, and that fossils are important clues to understanding its history and development. These facts seem obvious today, yet men for centuries refused to believe the evidence before their eyes that life had existed on earth for a very long time.

THE EARLIEST PALEONTOLOGISTS

The ancient Greeks recognized that marine shells found in outcroppings around the Mediterranean Sea marked areas that had once been under water. Herodotus, the Greek historian and traveler, mentioned fossil seashells he had seen in Egypt and drew the conclusion that the sea had at some time covered Lower Egypt.

But other Greek thinkers left behind some mischievous ideas, such as Aristotle's teachings that there had been only a single creation. These ideas became mingled with the Church's dogma of the literal creation in six days and effectively stifled men's sense of inquiry until the 15th century.

Leonardo da Vinci (1452–1519), the great painter and architect, reasoned rightly that the presence of fossils uncovered in Lombardy indicated that northern Italy had been repeatedly inundated by the sea. However, he was one of the few free spirits of his age, an age when most men described fossils as sports of nature *(lusi naturae)*, as seeds of life that had grown by accident in the rocks, or as the bones of unicorns, legendary giants, or sinners drowned in the Biblical Flood.

But it was in this age that fossils got their name. Georgius Agricola (1494–1555), the first professional mineralogist, used the Latin word *fossilis* in his great work on systematic mineralogy, *De natura fossilium*, published in 1546. To Agricola, *fossilis* meant anything dug up from the earth; later writers picked up the word from him and distilled its meaning to include only what we now call fossils.

The Renaissance brought an intellectual climate that allowed a more rational view of nature. Count George Louis Leclerc Buffon (1707–1788) recognized that the rocks of the Paris basin were the product of untold ages of erosion and deposition. He also saw that they could not be sandwiched into the estimate of the age of the earth, made from the Biblical chronology, of about 6,000 years. His contemporary, the Scottish geologist James Hutton (1726–1797), went a giant step further. While such men as Buffon were attempting to explain away the contradiction by presuming that catastrophes had hastened the changes they plainly saw, or else fancied that what the Bible spoke of as days should be read as epochs of time, Hutton boldly faced the facts. He put a solid foundation under the study of geology and of fossils with his principle of uniformitarianism—that geological forces in the past were like those operating today and that no extraordinary forces or events need be invented or conjured up to explain geological history.

Another Frenchman, Jean Baptiste Pierre Antoine de Monet, Chevalier de Lamarck (1744–1829), in his little book *Hydrogéologie* (1802), first definitely restricted the word "fossil" to "the still-recognizable remains of organized bodies." Elsewhere he made the first major study of invertebrate paleontology and advanced the theory that all life had developed progressively from rudimentary forms to its culmination in man—a theory of evolution.

At the same time, Baron Georges Chrétien Léopold Dagobert Cuvier (1769–1832) was laying the foundations of vertebrate paleontology. He had also discovered that a series of deposits made up the Paris basin and that each one had characteristic fossils by which it could be recognized wherever it appeared. This principle was confirmed by the independent discoveries made across the Channel by William Smith (1769–1839,) who is usually described as the father of English geology. Smith, a surveyor, examined the fossils found as his men dug canals, noted that he could use them as indicators of the type of ground to be expected elsewhere, and from this experience developed toward the end of the eighteenth century his fundamental principle of correlation—that like rock strata have like fossils by which the strata can be recognized wherever they are found.

With the work of such men, with the summing up of their discoveries by Sir Charles Lyell in his *Principles of Geology,* and with acceptance of Darwin's theory of evolution, men finally began to comprehend the meaning of fossils.

HOW FOSSILS ARE FORMED

Becoming a fossil is no easy adventure. In the more than two billion years of life on earth, an incalculable number of organisms have lived and died. If all had been preserved, our earth would have become nothing but a mass of fossils. Fortunately, most organisms have returned to the earth from which they came and left no fossil litter behind.

A creature destined to become a fossil usually is one that possesses hard parts, such as a shell, horny armor like a crab, or bones that will resist the abrasive effect of water and wind and the appetites of bacteria. As has been said, "You have to be hard and tough to get into the fossil record." Besides being tough the creature must come to rest in a place where it stands a good chance of being buried before it decays or disintegrates. If it is not buried deeply and quickly, aerobic bacteria will reduce it to rubble; or water, given enough time, will dissolve it.

For this reason, fossils of some kinds of organisms are rarer than others. Butterflies, for example, are common in nature, but their soft bodies and fragile wings leave few epitaphs in nature's graveyard. The soft tissues of jellyfish likewise stand little chance of becoming fossils even though their marine environment is far more favorable for this purpose then land.

The ideal place to become a fossil is at the bottom of a quiet sea or lake where the prospective fossil is safe from damage and where it is covered rapidly with fine sediment. Clay is excellent. The sediment protects the tissues and helps to exclude predators and solvent water. If the water is poisoned with dissolved chemicals that will keep predators away, so much the better for the future fossil's chances.

Consequently, fossils are most commonly found in fine-grained sedimentary rocks, such as shale derived from the compressed clay and silt of an ancient sea or lake bottom, or in limestone formed in warm sea water by chemical precipitation and the constant accumulation of carbonate shells of living organisms. Wave currents strong enough to wash in sand and gravel are also strong enough to sweep away or damage most shells and skeletons; hence only the toughest fragments of fossils are generally present in sandstones and conglomerates.

Fossils, then, are not only fairly rare as compared with the plenitude of life through the long history of the earth, but they also give a distorted view of it, because of nature's favoritism to certain types of organisms. Furthermore, comparatively few of earth's fossil resources have been tapped. It has been said that all the fossils available to science represent the variety of life of the past about as accurately as one mosquito represents the enormous variety of insect life today. Reconstructing the past from fossils is like trying to recreate the Parthenon from a basket of marble fragments. Here a piece of column, there a tile from the roof, here a limb of a statue—how did they once fit together?

But imperfect though the fossil record may be, it is the definitive record written in the rocks, and it is written in many ways. Fossils can be divided into half a dozen categories of preservation. Most specimens found by the average collector will fall into two or three categories, but some acquaintance with the others is also his legitimate concern. These divisions, in order of progressive change, include fossils preserved by freezing, drying, original preservation, petrifaction, and carbonization, and those preserved as casts and molds.

The person who brought this specimen to a museum identified it as a fossil cow's head. It is a piece of flint, a pseudofossil.

Freezing

The best-preserved fossils are those of organisms that have been frozen quickly. Only a few species of not-very-old fossils have been so preserved to this day, notably some of the large Pleistocene mammoths that were mysteriously frozen while wandering about Siberia and Alaska about ten thousand years ago. These mammoths, still melting out of the permafrost,

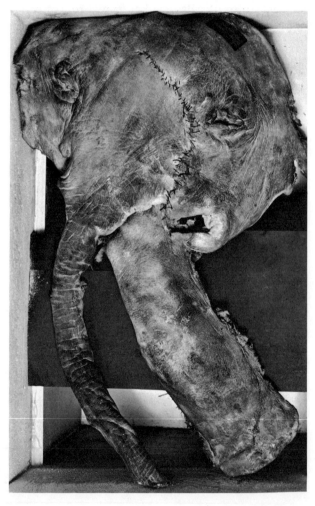

Frozen mammoth from Alaska in refrigerator case. (Photo American Museum of Natural History, New York)

were frozen so rapidly that the last mouthful a pachyderm was munching before its unfortunate accident remains unchewed in its mouth. Such fossils are difficult to collect and even more difficult to display. The American Museum of Natural History in New York has the front part of a small mammoth from Alaska on display in a glass-walled freezer. Similar frozen fossils are eagerly collected by the Eskimos, who chop them up and feed them to their dogs, a strange use indeed for fossils.

Drying

About the time the mammoths were freezing in the northlands, other vertebrates crawled into caves in southern desert regions and died. In this aseptic environment they became mummified. Bones and tissues of these desiccated denizens of the desert are preserved, although often they fall apart at the slightest touch. Even skin and hair retain their original color. Such fossils are the only accurate evidence available to the scientist trying to restore a bag of bones and give it the proper clothing.

Original Preservation

Bones, teeth, shells, and wood can be buried and remain chemically unchanged for millions of years. Most Miocene, Eocene, and Pleistocene shells such as those found in Maryland, Virginia, Florida, and California are essentially the same as when they were buried. Often the only clue to tell these 20-million-year-old shells from their modern counterparts is loss of color and luster. Many bones dredged up in midwestern gravel pits are little changed since they once held together ice-age animals.

One location in Tennessee produces 135-million-year-old Cretaceous shells of remarkably modern appearance, even to the pearly luster. Sometimes the original aragonite of the shells has changed to calcite, chemically the same but different in crystal structure. Logs embedded in Eocene coal deposits often resemble a modern-day fireplace log and are quite capable of being burned.

Relatively recent fossils of animals preserving flesh, skin, and hair have been dredged from peat bogs, where tannic acid in the water has saved them from decay. The body of a man so fossilized rests in the science museum in Copenhagen, Denmark, his face clearly showing a three-day stubble of beard. In Galicia, now part of the Soviet Union, a rhinoceros carcass was found pickled in an oil seep.

Petrifaction

No category of fossil preservation is so misunderstood as petrifaction (sometimes spelled petrification). Everyone has heard of petrified wood. The word "petrified" comes from the Greek word *petros,* meaning "stone," and petrifaction literally means "turned to stone." Unfortunately, many persons consider any fossil petrified. But strictly speaking, a fossil is petrified only when additional minerals have been deposited in pores or cavities in the fossil, or when the fossil is entirely replaced by other material. Consider a piece of wood. It can petrify in three distinct ways, each with a distinctive result with a distinctive name:

1. By filling the empty spaces with some mineral, as water fills the empty spaces in a sponge. This is called *permineralization.* Dissolve this mineral, and the original piece of wood remains.
2. By filling the empty spaces with mineral, then dissolving the cellulose and wood fibers and replacing them with mineral matter, often of a different color. The result is a piece of stone that faithfully reproduces

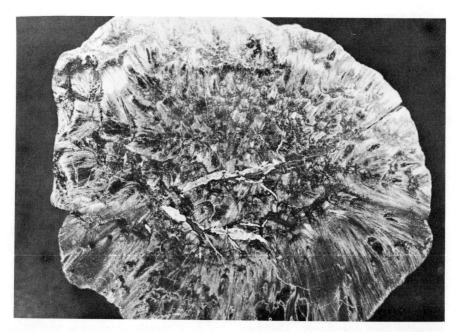

Petrified wood first became common in the Pennsylvanian period. The lack of growth rings on this *Sigillaria* trunk suggests an even climate with no seasons to affect rate of growth. The wood is replaced by calcite. Knoxville, Iowa.

Wood replaced by quartz, found in Utah. Growth rings are indicated by alternating brown and white quartz. Cracks filled by clear quartz suggest wood dried before silicification.

every cell and detail. But dissolve away the mineral matter and there is nothing. This is called *histometabasis*, or more commonly, *replacement*.

3. By surrounding the wood with mud or sand that hardens around it. When the wood decays or is dissolved, a mold is left which fills with

Once driftwood, now quartz. This piece of petrified wood from New Mexico floated long enough to wear away bark and soft wood before sinking and becoming a fossil.

The wood was replaced by barite, which has crystallized into typical radiating masses. Orlando, Oklahoma.

mineral matter. The result is a piece of what looks like wood on the outside, but inside may be banded agate or even a geode with sparkling amethyst crystals, showing no cellular detail at all. This type of replacement is often called *pseudomorph* ("false form"). This also occurs in minerals where one mineral has faithfully replaced another mineral crystal, such as a calcite crystal replaced by quartz that retains the crystal shape of the calcite.

All three of these quite different processes are correctly termed petrification or petrifaction.

PERMINERALIZATION

Bone, plant materials, and many shells are porous enough for permineralization or replacement to occur. The best example is the coal ball, found in some coal mines. This is nothing more than a mass of Coal Age plant fragments and seeds that has become permineralized by calcite or sometimes by iron sulfide (pyrite or marcasite). When the coal ball has been permineralized by calcite, the collector can perform the interesting opera-

Chonetes fragilis, a brachiopod from the Devonian shale of Sylvania, Ohio. Like most brachiopods from this area, it is replaced by pyrite.

tion of peeling off a thin layer of the actual woody material of these 275-million-year-old plants. (Details are explained in Chapter 10).

Permineralization is common in petrified wood, but too often the mineral filling the empty spaces is quartz, which can be readily dissolved only by hydrofluoric acid. Peels, however, have been made of silicified wood in the same way coal-ball peels are made. Quartz and calcite are

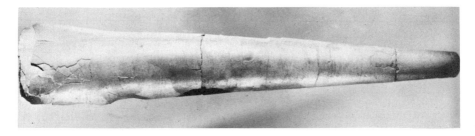

The small straight cephalopod *Pseudorthoceras knoxense* is often found brightly pyritized like this one in Pennsylvanian black shales. Omaha, Nebraska.

the minerals most commonly found in permineralized fossils, but others include pyrite, marcasite, barite, selenite, opal, and manganese oxide.

REPLACEMENT
Replacement is often met with. Coal mines sometimes produce beautifully "pyritized" snails, clams, and brachiopods. Actually, most of this "pyrite" is marcasite, the less stable form of iron sulfide. It is hard to draw the line between a replacement and a pseudomorph, especially in fossil shells, although, strictly speaking, in replacement only the shell is replaced; in a pseudomorph both the shell and its filling are replaced. Beautifully pyritized brachiopods have been found in the Silica shale (Devonian) of Ohio. Silica-replaced brachiopods occur in New Mexico, and silica-replaced corals, stained a beautiful red, are found in Utah. Large colonies of *Lithostrotionella*, a Mississippian coral replaced by colorful quartz, are found in southeast Iowa. These retain the fine detail of the original organism. Another colonial coral, *Hexagonaria*, occurs as handsome specimens filled with calcite in Devonian formations in Michigan. When washed up, rounded by the waves, on Lake Michigan beaches, these are

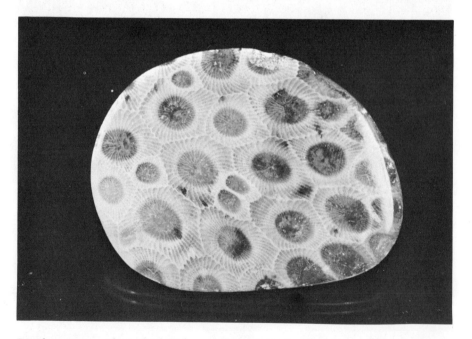

Petoskey stone, perhaps the best-known invertebrate fossil, the state rock of Michigan. This Devonian coral, *Hexagonaria*, is replaced by calcite in contrasting colors. It takes a fine polish, as in this piece.

known as "Petoskey stones" and are eagerly collected. This soft, tan material, which polishes well to display the inner workings of the fossil, has been declared the official state rock of Michigan.

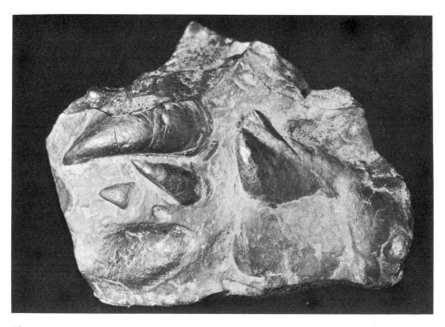

The original matrix, limestone, as well as these clams, is replaced by hematite. Such dark-brown fossils are common in the Minnesota iron range; however, this one is from Missouri.

PSEUDOMORPHS

True pseudomorph plants are found in lava flows in the Pacific Northwest, where trees engulfed by hot lava were burned away but left a faithful mold in the hardened lava. Even a mold of an unfortunate rhinoceros, formed in this fiery furnace, is known from there. Such cavities may become filled with agate. A slab of this agate does not look like a slab of petrified wood from the Petrified Forest in Arizona; it looks like any banded agate because it had a chance to replace only the shape of the tree, not the grain of the wood.

Agate pseudomorphs of coral are dug from the bottom of Hillsborough Bay at Tampa, Florida. They show traces of coral structure on the outside but the interior cavity is lined with vividly colored layers of chal-

Silicified fossils may be replaced with transparent or translucent quartz that allows light to pass through the fossil, such as this fossil snail from the Miocene of Florida.

cedony or with drusy quartz crystals; the original coral had completely dissolved before the chalcedony arrived.

HOW PETRIFACTION OCCURS

The process of petrifaction can most easily be understood by analogy to a water pipe in a house. Most water has some degree of "hardness"— that is, it contains minerals it dissolved while percolating through rock and soil underground. Given enough time, water will dissolve almost any mineral. The process is hastened when water absorbs carbon dioxide gas and becomes a weak acid. It picks up carbon dioxide while falling through the air as rain, or by release of the gas from decaying organisms in the soil. Furthermore, water draining slowly through soil will pick up such organic ammunition as tannic or humic acid. This acid water seeps slowly into the ground, where it dissolves minerals in its underground passage.

Water can hold only so much dissolved mineral matter before it is saturated. As hot saturated water cools or evaporates, the excess mineral matter crystallizes on any object in contact with the water. This may be a water pipe. Each surge of water passing through the pipe deposits an imperceptible layer of mineral. This is especially true in hot-water pipes. Sometimes pipes become so filled with mineral deposits that the plumbing must be replaced.

Wood and bone, and to a lesser extent shells, are composed of closely connected pipes through which water can move. When these organic remains are buried under a lake or ocean, water has a chance to move through these natural pipes and to deposit layer after layer of mineral in the channels or cells. The cell filling may become brightly colored by tiny amounts of dissolved metals; the bright reds, yellow, greens, and browns in petrified wood are caused by iron. Blacks are often made by trace amounts of manganese, and copper or nickel can create greens. Little by little the tubes are plugged up, and the result is a permineralized fossil. If a mold is filled, it is a pseudomorph.

Petrified wood and bone are commonly filled by quartz-bearing solutions, especially when the fossil is buried in volcanic ash, which is rich in silica. Water percolating through the ash, which may remain warm for a long time, dissolves the silica and immediately deposits it in the wood or bone fibers. Whether this takes place in a relatively short time is uncertain. Nobody has ever been able to petrify wood effectively in a laboratory.

Carbonization

Carbonization, also known as distillation, is one process that preserves fossils of soft-bodied animals and leaves and stems of plants. Carbonization chemically alters the proteins and cellulose of tissues through degradation by bacteria, by chemical action, and by pressure and heat, until only carbon films remain. Other organic materials are dissipated as gases —carbon dioxide, methane, hydrogen sulfide, and water vapor. For example, a thick, fleshy fern leaf of Pennsylvanian times falls into mud. Hundreds of millions of years later a paper-thin black carbon copy of the leaf is found, perfectly preserving its details in shale. Coal is formed in the same way, but on a much larger scale. Carbonized plants are common in the shale overlying coal seams.

Carbonization preserved specimens of the Silurian worm *Lecthyalus gregarius* which wriggled about Chicago seas 400 million years ago. Carbonized fossils are by no means confined to such ancient rocks. The Cretaceous and younger formations of Tennessee contain plant leaves of this type, and so do the ash beds at Florissant, Colorado, the shales at Green River, Wyoming, and the Latah formations near Spokane, Washington. More commonly, however, Cenozoic plants exist as impressions.

The original plant material of this fern has been reduced to a carbon film. This is typical of plant fossils found near coal seams. Pennsylvanian period; Terre Haute, Indiana.

Soft-bodied animals, such as this worm, are rarely preserved except as a carbon film. Parts of this Silurian annelid, *Lecthyalus gregarius*, found near Chicago, are still attached to the mold in the limestone.

These ants were trapped in resin which hardened into amber over the last 40 million years. They are thin, carbonized films with hollow centers. Poland.

Casts and Molds

An organism, such as a shell, is buried in sediment, such as the ooze on the bottom of a sea. Water later dissolves the shell, leaving a hollow in the hardening sediment. The sediment becomes rock enclosing a cavity that exactly preserves the external pattern and shape of the shell. This is a *mold*.

Sometimes a mold becomes filled with sediment or other mineral matter, producing an exact duplicate of the external pattern and shape of the shell. This is a *cast* from the mold. The pseudomorphs discussed under petrifaction are casts. The petrified wood pseudomorphs found in western lava flows are for this reason commonly called limb casts.

Dinosaur and bird footprints are molds in the sediments that yielded to their footsteps long ago and then hardened. A rock layer that formed above and filled them would, when pried apart from the lower rock, show a cast of the footprint.

In some Paleozoic rocks, notably the Silurian Niagaran dolomites in the Middle West, unusual double molds are encountered in fossils of brachiopods, crinoids, snails, and trilobites. When a slab is broken open, a mold is found that shows the external details of the creature. Rattling

Mold of a trilobite, *Calymene niagarensis*, in dolomite of Silurian age. This mold was empty; no steinkern was found inside. Racine, Wisconsin.

This odd fossil is scarcely recognizable as a cup coral. This is a mold of the top of the coral, like a piece of clay pressed into the hollow top and then removed.

Sandstone blocks from an Ohio strip mine were found to contain tracks of an early amphibian. (Photo by Betty Crawford)

Tracks of the Pennsylvanian amphibian found by Drs. Dick Patterson and Dwayne Stone of Marietta College. (Photo by Betty Crawford)

Many early Paleozoic snails, such as this one, are found only as internal molds. Such a snail may have been dull on the outside, or it may have been highly ornamented, but this fossil represented only the living chamber within the shell. Devonian; Michigan. (Photo Michigan Conservation Department)

External mold of a straight cephalopod. Silurian; Illinois.

A drawing of *Glycymeris*, a clam from the Miocene of Florida. At left is a shell in perfect preservation, a right a steinkern of a similar shell. A steinkern is a cast of the interior of the shell. (Drawing by Betty Crawford)

inside the mold is a loose fossil that has a superficial resemblance to the mold in the rock. This is an internal mold, a mold of the inside of the shell, or it can be thought of as a cast of the inside of the shell. Paleontologists refer to these double-duty fossils by their German name, *steinkerns* ("stone kernels").

Such double molds are formed by an elaborate maneuver of nature. A clam, for example, is buried and becomes filled with mud. If the shell is dissolved away, an external mold would be left. Inside it would be the internal mold of mud, taking the place of the body of the clam and faithfully reflecting the detail of the inside of the shell, just as the external mold reflects the quite different detail of the outside of the shell.

Such a pair of molds is far more useful than an external mold alone, because it shows not only external details, but such important things as muscle attachments in mollusks, body-cavity shape and size in crinoids, and thickness of the horny exoskeleton in trilobites.

The remarkable soft-bodied animals found near Chicago and discussed in Chapter 1 are casts and molds. Insects in amber are molds; the delicate insects long ago turned to dust and gas, leaving behind a cavity lined with carbon so detailed that even the smallest parts of the antennae are visible. A few dinosaurs and fish have been found with the bones in place and an impression in the shale of the fleshy body of the animal itself. Evidently the mud became hard enough before the tissues decayed to hold the impression made by the bulky body as it sank in the ooze. To a like circumstance we owe preservation of fine detail of the bodies

A steinkern, or internal mold of *Calymene*. Silurian period; Lemont, Illinois.

of the large marine ichthyosaurs in the Jurassic rocks at Holzmaden, Germany. Even the soft fleshy paddles of the reptiles made their mark in the rock.

Archaeopteryx, a Jurassic link between reptile and bird, had never been recognized as a bird until specimens came to light in the limestone quarries near Eichstatt, Germany. The bones were in place on the slabs, and so were distinct impressions (molds) of the feathers, wings, and tail of this early toothed bird.

MISCELLANEOUS FOSSIL FORMS

Fossils include not only the remains of actual living creatures but all those bits and pieces that show evidence of the existence of such creatures. Among these miscellaneous forms are trace fossils, borings, and coprolites. Other fossil forms are organic structures composed of algae or bacteria.

Trace Fossils

These include trails and burrows and other fossil evidence of the activities of once-living creatures. A tidal mud flat at low tide is an amazing sight, covered with innumerable trails left by clams, crabs, and worms and speckled with holes that are the tops of burrows of a multitude of worms and clams. Should this area dry sufficiently to harden, the next tide might not destroy these markings but instead gently cover them with a layer of mud. When turned into rock and split apart, these layers would reveal the trails and burrows of extinct organisms. Some of the oldest known fossils are burrows and trails, evidence that something crawled in Pre-Cambrian seas but never was preserved as a fossil.

Occasionally the nature of such a trail becomes clear through the lucky discovery of its fossilized creator, too. Jurassic horseshoe crabs have been found in Germany in this situation at the end of their last crawl.

In the early days of paleontology, ridgelike and tubelike swellings in rocks that showed definite indication of having been formed by something living were classified as marine algae. Because many showed definite

A variety of trails, burrows, and resting places are the only fossils in this thin sandstone.

branching, this marine-algae theory gained support. Many generic names still end in -phycus, such as Arthrophycus or Dolatophycus, signifying algal origin. In the same way, such names as Fucoides, Algacites, and Chondrites cling to remains not now believed to have any connection with the plant kingdom.

Gradually these trace fossils became accepted as burrows and trails after their resemblance to those made by modern worms and pelecypods was noticed. Many are believed to be subsurface burrows that filled with a different sediment. This sediment now appears in relief on the surface of a slab. A thin, buried layer of sediment of different texture often created a plane of weakness that offered an easy path for the burrowing creatures. It also offered an easy parting plane when the rock was split some millions of years later by a fossil collector.

A single individual may have made a series of quite different markings in the sea bottom, depending on whether it was crawling, running, feeding, or burrowing. Thus it is difficult to relate these problematical fossils to their creator. The system proposed by Adolf Seilacher for classification of these fossils seems to be the most reasonable. He has grouped trace fossils by their ecological similarity whether their proprietors were similar or not. A tubelike burrow made by a trilobite is similar to one made by a worm. They cannot be told apart unless a very dead worm or trilobite is found at the end of one. So this system has merit. His five basic groups are:

1. Dwelling burrows: Tunnels made as living quarters, usually at right angles to the bedding of the layers, originally opening out on the surface.
2. Feeding burrows: Tunnel systems, originally dug below the surface, excavated by sediment-eaters while searching for food.
3. Feeding trails: Tunnels and bands that are extremely winding and very numerous, probably made on the surface, rarely crossing each other, made by organisms also in pursuit of food or on their way somewhere.
4. Resting trails: Isolated impressions with a vague outline of their producer, probably representing resting spots or nap sites.
5. Crawling trails: Showing variable direction and imprints of legs, usually made on the surface.

Borings

Bored holes are quite common, particularly in shells, both modern and fossil. Some are the work of predatory snails that rasp a tapering hole through the shell to get at the delicate meal inside. Borings are made in

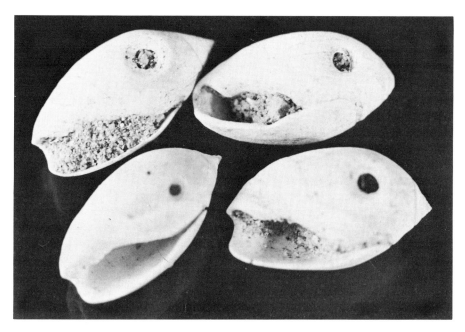

These California snails show neat holes bored by a cannibalistic cousin to reach the meal inside. Pleistocene.

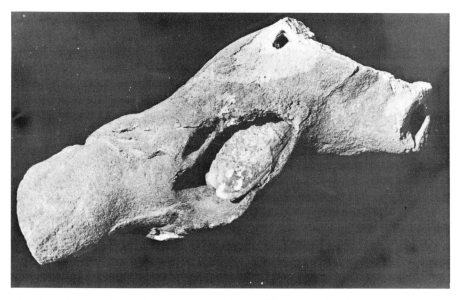

The clam that bored into this ancient Florida coral became a fossil while resting in its burrow. Miocene; Tampa, Florida.

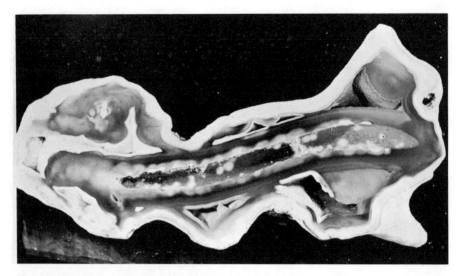

The tube inside this silicified Miocene coral from Florida was bored into the living coral by a clam similar to the teredo.

living and dead shells (also corals and even solid rock) by barnacles, which leave behind a characteristic sac-shaped hole, often very small. Modern shells worked over by barnacles are common on Florida beaches. Worms leave tiny, narrow tubular borings that may be curved or branch-

Coal is not often thought of as a source of recognizable fossils. This Eocene fish was found in a coal seam near Frankfurt, Germany.

ing. Small sponges have left similar borings, often radiating from some central point. Bryozoans and boring corals leave tiny tubelike holes, again most commonly in shells.

Everyone is familiar with teredo wood, which has become the official state rock of North Dakota. The teredo (or shipworm) is just as active today as it has been ever since it first made its appearance in the fossil record in the Jurassic period. The teredo is not a worm but a clam turned lumberjack. It is a member of the pelecypod (clam) suborder *Adesmacea*, comprising clams that bore. The teredo drills into wood, quickly riddling unprotected wooden piling, docks, and ship hulls. Teredo-bored logs are not uncommon in petrified wood, and occasionally a piece is found with the teredo still in place, its burrow, like itself, turned to solid agate.

Coprolites

Coprolites are fossil excrement of anything from a mighty dinosaur to a fish or worm. Fish and shark coprolites are of particular interest because they often preserve tiny scales and teeth that reveal what the predator ate and what lived in the area where it dined. Such teeth and scales sometimes represent fish never found as fossils elsewhere.

A fish coprolite nearly filling a concretion. Small bones, scales, and teeth from past meals are sometimes found in coprolites. Pennsylvanian; Braidwood, Illinois.

Such unmistakable coprolites of mammal origin are rare, but are found in a Washington locality.

Organic Structures

Pre-Cambrian swellings that have a layered appearance like a giant biscuit apparently were formed by algae that removed carbon dioxide from lime-rich water, causing precipitation of calcium carbonate. The vast iron deposits of Minnesota are believed to owe their origin to bacteria that took iron out of solution in the water and carried it to the bottom as an oxide when they died. For this reason, these iron-rich sedimentary layers may be considered a gigantic fossil graveyard.

III

THE NAMING OF FOSSILS

Understanding the nature of fossils is the first step; recognizing them by their names is the second. This second step requires acquaintance with the system for classifying living things and once-living things devised by a Swedish naturalist, Carl von Linne. He perfected his system over many years and published the tenth edition of his book, *Systema Naturae*, in 1758. This is the basis of zoological nomenclature.

CLASSIFYING PLANTS AND ANIMALS

In what is known as the Linnaean system, each organism—animal or plant, living or fossil—is identified by two names. The first is the generic or general name (the name of the genus), which is written with a capital letter; the second is the trivial name, which is not capitalized. Latin and Greek words are adapted to the purpose of naming organisms so that the names will be identical in all countries and all languages. Thus man is *Homo sapiens;* the common cat is *Felis domestica;* and the common oyster *Ostrea virginica*. The two words, which together form the species or specific name, are printed in italics. The generic name can be thought of as similar to the surname, such as Smith or Doe, and the trivial name to the baptismal or personal name, such as Marcia or John.

The generic name can be used only once within a related group of organisms, but the trivial name can be used with other generic names. For instance, the trivial name "robustus," of obvious meaning, is applied to some two dozen genera (the plural of "genus") of invertebrate fossils. At least four of the genera—*Scaphiocrinus, Zeacrinus, Protaxicrinus,* and *Synbathocrinus*—are crinoids. But a generic name such as *Zeacrinus* could not be used for any other animal. In practice, generic names are not duplicated. They must not be duplicated within a kingdom—that is, they cannot be used for two animals or two plants.

An organism retains its trivial name even though reassignment to a different genus is necessary if the fossil classification is reorganized when better information about relationships becomes available. For a full description, the name of the person who originally described and named the fossil and occasionally the date when he did so are added to the generic and trivial names. A common Pennsylvania-period snail from the Mississippi Valley, for example, is called *Worthenia tabulata* (Conrad). Conrad named it *Turbo tabulatus,* but research later showed that it belonged to the genus *Worthenia.* The trivial name was retained, though changed in grammatical gender to agree with the generic name, and Conrad's name was put in parentheses to show that the species has been assigned to a different genus.

A third name, that of a subspecies or variety, may be added to the two terms forming the specific name. A subspecies or variety is a form that is not sufficiently different, in the classifier's judgment, to stand as a distinct species, yet is morphologically distinct within the character range of the species.

A next largest unit above the genus is the family. Under generally accepted rules, names of families are now formed by adding -*idae* to the stem of the name of the genus chosen as the type for the family. Man, for example, belongs to the family *Hominidae,* and the cat to the family *Felidae.* Subfamilies (smaller groups) have names ending in -*inae,* and superfamilies (larger groups) have names ending in -*acea.* Larger groupings, in ascending order of size, are the orders, the class, and the phylum. Names in these three largest classifications are capitalized but not italicized. Detailed description of the phyla of interest to the fossil collector has been postponed until the last chapter of this book, a place more convenient for reference in identifying and classifying a puzzling specimen.

The Holotype

Classification is based on recognition of similarities. The species, the category of classification immediately below the genus, comprises organisms that are similar to one another and are capable of interbreeding and producing fertile offspring. In order that other paleontologists will have a concrete example for recognition of the characteristics of the species, the person who first describes and names it designates a specimen as the *holotype*, or sometimes he designates a group of like specimens as a type series, including one individual specimen as the holotype. This is the standard to which collectors and classifiers can refer, the keystone of the species.

Genus is an abstract concept by which classifiers can associate species in which they have reason to recognize biological relationships and similarities. The author of a genus designates one of the species included in his genus as the *genotype*, which then becomes the standard for reference to that genus. The cat, for example, is placed in the same genus as the lion, tiger, and cougar, while each belongs to its own species. Like the genus, the larger units—family, class, and phylum—are synthetic groupings based upon more and more general resemblances, just as the human family is included in the nation and the race.

A recent example affords glimpses of paleontologists at work studying and classifying a fossil. Near Bishop, California, 31 specimens of a fossil resembling the present-day sea cucumber were collected in Lower Cambrian shale. The organism was associated with fellow members of the echinoderms, to which crinoids and starfish belong, and with trilobites. Unlike the typical crinoid, however, it apparently was not held to the

How a variety, or subspecies, differs from its parent species is not immediately apparent. At left is *Phacops rana*; Devonian; Alden, New York. At right is *Phacops rana milleri*; Devonian; Sylvania, Ohio. They differ mainly in the number of facets in the compound eyes.

Helicoplacus gilberti, newly discovered echinoderm. (Photo courtesy Professor K. E. Caster)

bottom by a stalk but was free-living. Its closest relative in the fossil world appeared to be a group of starfishlike animals known as edrioasteroids.

What was unique about this new organism was the spiral arrangement of the tiny hard plates of its body. Later it was deduced after close study that the organism could expand and contract its armored skin.

Two species were distinguished—*Helicoplacus gilberti* and *Helicoplacus curtisi*, from the Greek *helix* and *plakos*, meaning "spiral plates"—and it was proposed to place them in a new class, the *Helicoplacoidea*, in recognition of their major differences from already classified fossils. In their announcement, Professor J. W. Durham of the University of California at Berkeley and Professor K. E. Caster of the University of Cincinnati asserted that from the discovery "it is apparent that considerable doubt is cast upon the generally accepted view that the ancestral echinoderm was attached. Need for a reconsideration of all subphyla of the Echinodermata hitherto proposed is also indicated." Holotypes of the two species were deposited in the University of California museum.

Convergence

Fossils were at one time classified primarily by similarities of structure of the hard parts of organisms. This was found to be misleading as paleontologists discovered that some entirely unrelated creatures had developed similar structures, presumably because they were useful to different organisms that had adopted similar ways of life. This trick of nature, which Professor George Gaylord Simpson has called the "bane of the taxonomist," is called *convergence*.

Perhaps the most obvious example is that intelligent animal the porpoise, which has evolved a body shape and structure first adopted by the ichthyosaur 100 million years ago. Size, shape, and placement of front flippers are similar. Yet the ichthyosaur was a reptile related to the dinosaur; the porpoise is a mammal, albeit a sea-dwelling one, related to man. If both were known only from their gross outlines, they might have been classified in the same family or even the same genus, when in reality they are not even of the same class.

Convergence of a type more potentially confusing to the paleontologist occurs in a case such as that of the horn corals, which are coelenterates, the rudistid clams, which are mollusks, and some brachiopods, all of which in some extinct species took the form of a cone attached by its small end to the sea bottom or some other anchor. Although they belong to three different phyla, they can be told apart only after close study.

The phenomenon of convergence is only one of the problems that classifiers of fossils face, working as they often must with the fragmentary or imperfect evidence that time has left them. But occasionally they have a stroke of luck. This happened to Dr. H. B. Fell of Victoria University, Wellington, New Zealand, in the course of a long study of the evolution of starfish. From the British museum he obtained part of a "living fossil," the dried arm of a sea star or somasteroid, an animal previously known only from fossils of 400 million years ago, when it was presumed to have become extinct.

From a detailed anatomical study of the tissues (not possible from fossils) and later of specimens caught alive in the Pacific off Nicaragua, he confirmed that the arms of somasteriods had structures like the arms of crinoids but that the body was flat and shaped like a starfish. He described the "living fossils" as "the oldest type of astrozoan (starfish) echinoderm yet discovered" and suggested that "the marked resemblance to crinoids now evident in somasteroids" made it probable that all star-shaped echinoderms had descended from crinoids.

Such geneological researches excite the spirit of discovery and tax the skills of professionals. They are beyond the reach of amateurs, but the thrill of finding a beautiful fossil by hard work and the pleasure of preparing it are the collector's reward. They are his school and his deepest pleasure.

Furthermore, he can enjoy his specimens for themselves, for fossils are often beautiful, like ancient sculptures or rare gems and minerals. Amber, pale and waxy, or vibrant with glowing hues of reddish brown, a showcase of ancient insects. Pyritized brachiopods and snails miraculously transformed to brassy gold, and the theatrical contrasts of pyritized starfish and trilobites on coal-black slates from Germany. The impudent faces of trilobites, like African carvings in ebony, peering from their tombs of 400 million years ago, and the bold splashes of color in an agatized log from Arizona.

Besides this aesthetic appeal, fossils have an appeal rooted deep in life itself. They once lived. To be able to hold a fossil in your hand, to imagine what sunrises and sunsets it saw eons ago in a world that we can only dimly imagine, is something that transcends ordinary experience. This is, as William Blake wrote:

> To see the world in a grain of sand,
>   And a Heaven in a wild flower;
> Hold infinity in the palm of your hand,
>   And eternity in an hour.

In a like mood, Professor Simpson wrote: "Our allotted span is a few years, and most of us can see with our own eyes only a minute part of the earth around us. But our minds need not be restricted to these narrow limits of time and space. They can range through the past and can see all the curious creatures and scenes of life's history in ever-changing sequence. A fair title for a book on paleontology might be, 'How to Live a Billion Years.' "

IV

WHERE FOSSILS OCCUR

Sedimentary rocks, particularly shales and limestones, are the storehouse of 99 percent of the world's fossils. One of the three major types of rocks that form the crust of the earth, these rocks are distinguished from the other two types because they are formed of sediments. Some are composed of silt, sand, and pebbles deposited mechanically by wind and water; some are chemical sediments precipitated from water or taken from water by plants and animals and then deposited with their bodies; and others are sediments that arise from combinations of these agents. Such rocks are commonly stratified or layered, and they are formed of materials and under conditions that are part of our everyday life.

The other two types of rocks—igneous and metamorphic—are the products of more deep-seated dynamic earth forces. Igneous rocks originate as molten magma. They may pour out on the surface as lava and harden, or they may slowly crystallize deep within the earth and then be exposed by erosion. Basalt is an example of volcanic igneous rock, and granite of crystalline igneous rock. Igneous rocks rarely contain fossils; granite never does.

Metamorphic rocks, the third type, are those formed from other rocks by heat and pressure. These agents change the structure of the rock, often recombining its constituent chemicals into new mineral species.

Volcanic ash may preserve fossils. While excavating solidified ash that covered Pompeii, archaeologists found strange cavities. Plaster forced into them made intricate casts of the bodies of men and animals. In this cast of a dog, details such as the collar are still visible. No bones were found.

Such a working-over is not a favorable situation for the preservation of fossils, although marbles, metamorphosed from fossiliferous limestone, often are cut for ornamental use to display fossil inclusions. Slate (metamorphosed shale) from Bundenbach, Germany, is noted for its spectacular pyritized starfish, trilobites, and other fossils.

SEDIMENTARY ROCKS

Sedimentary rocks, however, are the principal bank on which the fossil collector will draw. Knowledge of the principal types, the conditions under which they were formed, the relationships of the strata he will work in

Fossils in slate are often distorted. This trilobite, *Ogygopsis klotzi*, comes from Cambrian slates exposed on Mt. Stephen in British Columbia. It is flattened but shows little distortion otherwise.

the field, and the fossils he might expect to find in a specific sequence of sedimentary rocks is, therefore, part of the capital he will need if he is to make profitable use of his efforts.

Conglomerate

The coarsest-grained sedimentary rock is called conglomerate. It contains particles larger than sand, although it may contain finer particles, too. It looks very much like concrete. One variety is called pudding stone, from its resemblance to a pudding studded with fruit. In fact, the gastronomy-minded French use the word *poudingue* for conglomerate. The pebbles and rock fragments are usually cemented together by calcite or quartz. If the rock is made up of rounded pebbles, it is called conglomerate; if the pieces are broken and angular, it is breccia.

Waves and currents strong enough to drag in pebbles and whisk away sand and silt are also strong enough to break up potential fossil material that becomes mixed with the gravel. A conglomerate layer formed in this way is usually poor hunting ground for fossils, although it may contain pieces of petrified wood, bone, coral, or tough shells. But while crossing areas where conglomerate occurs, look for fossils in the pebbles themselves; you may make a rich haul.

Coarse-textured conglomerate is a poor medium for preserving fossils.

Sandstone

Sandstone is just that, stone formed of cemented sand. It is usually too coarse to preserve delicate fossils, but many leaves lurk in the Dakota sandstone of Montana, and hundreds of brachiopods in the Oriskany sandstone of New York. Foraminifera and ostracods, little fossils that are themselves the size of sand grains, are abundant in marine sandstones of the Cenozoic era.

Some marine sandstones include poorly preserved fossil shell molds. Water moved freely through these sands before they became stone, dissolved the shells contained in the sand, and left behind a grainy mold.

Sandstone can also be the hardened relic of a sandy riverbank or lake shore that existed millions of years ago. In Connecticut and Texas, dinosaurs ambled across these ancient shores and left behind their ample footprints. Some even left their mighty bones, too, and shifting river channels covered them with sediment. Such deposits of dinosaur bones are found in many western states. Heedless of such company, worms and bottom dwellers crawled on and in the sandy bottom, leaving behind their own fossil trails.

Shallow ocean bays that existed 275 million years ago collected quanti-

Fossils are usually poorly preserved in sandstone, and identification is difficult. These clams are from Kansas.

ties of Coal Age plants, torn loose from swamps near the shore during flood stages and floated out into the bay. There they became buried in sand. This occurred so rapidly in some areas, such as near Ottawa, Kansas, that the plants are found buried upright or at odd angles in the thick sandstone. The plants are preserved as carbon films, detailed enough so

This crab is an example of a well-preserved fossil found in sandstone. Oligocene; Washington.

that the genus of the plant can be identified, but not in as faithful detail as fossils of like plants preserved in the finer-grained shale just a few feet below the sandstone. Large upright logs and stumps were uncovered in Indiana in the last century where sandstone was mined to make millstones.

Shale

Shale is composed of clay, silt, or mud—materials smaller in size than the sand and pebbles of sandstone and conglomerate. Clay is made up of microscopic particles of aluminum silicates, such as mica, and of clay minerals such as kaolin, with fine particles of feldspar, quartz, and iron oxides. Shale cemented by calcium carbonate is appropriately called limy shale, but if there is more calcium carbonate than clay, it is shaly lime-stone. Both will fizz when touched by a drop of hydrochloric acid, but pure shales will not.

Black shale was formed where organic muds accumulated on the bottom of quiet waters like the Everglades, the Okefenokee swamp in Georgia, and some northern lakes. The bulk of the shale is composed of coallike rotted particles of plants and animals with a small amount of clay as a binder, now hardened into rock.

The muddy bottoms of early seas were home to brachiopods, clams, crinoids, and trilobites; almost every type of fossil can be found in marine shales. The crinoids of Crawfordsville, Indiana, lived and died and were buried in such deposits. So were the Cambrian trilobites of southern Utah and California: *Elrathia kingi* from a Utah location is common in collections. The handsome black *Phacops rana* of Sylvania, Ohio, is collected from a shale seam. The fine grain of clays preserves the structure of fleshy parts of soft animals as no other rock can. The unique Cambrian fossils of soft animals found in 1910 by C. D. Walcott in Canada were carbon films in the slaty black Burgess shale. A fine-grained shale is the matrix of the concretions that yield the Tully monster and other members of the Essex fauna of the Illinois strip mines.

Plant fossils are commonly found in shales. Carpeting the coal seams of the Illinois basin, the Iowa fields, and the eastern United States is a layer of shale formed, so geologists believe, where mud was washed in on top of thick layers of fallen vegetation. This must have occurred as the area sank slowly, drowning and killing the plants and turning the forest into a shallow, brackish lagoon or bay. Leaves, seeds, branches, cones, and trunks of the drowning trees and plants were buried and later carbonized. They are so well preserved that even the hairs that grew on

A road cut near Florissant, Colorado, cuts through thick layers of shale formed of volcanic ash. It contains fossil leaves from top to bottom.

the leaves of *Neuropteris scheuchzeri,* a seed-bearing fern, can be measured and counted.

At Puryear, Tennessee, clay shales are quarried that contain quantities of carbon-coated impressions of leaves that look like those from modern trees. This was probably a lake deposit, since a leaf will not drift very far in a stream before decaying or disintegrating.

The extensive Green River formation in Wyoming, outcropping in hills over a large area, was once the site of an Eocene lake where fish swam.

Fish sank to the bottom and were preserved in Eocene shales of the Green River formation, Kemmerer, Wyoming. Scales and bones have been lost from the front half where the specimen was weathered.

Killed by unknown causes, many thousands periodically sank to the bottom and were buried in a mud formed of volcanic dust. The water of this lake must have become poisonous, because the fish show no sign that they were killed by predators that certainly were present in the lake.

Shales, then, are an important source of fossils. Freshly quarried shales laid down in quiet waters may yield specimens in which all the delicate detail of the living animal is preserved. In weathered shale, however, nature leaves only the hard parts for the collector.

The fish-bearing shales are easily pried from hillsides west of Kemmerer, Wyoming.

Limestone weathers slowly, shale rapidly, so that the limestone in a road cut juts out over the shale layers, which are often rich in fossils. Massive limestones such as these are usually poor in fossils.

Limestone

Limestone is exceedingly common, is exposed over a wide area, and is abundantly fossiliferous. It has the further virtue of preserving its fossils

A limestone composed principally of one type of fossil may be named after that fossil; an example is this crinoidal limestone. Pennsylvanian; San Saba, Texas.

with little crushing and in fine detail. Limestone is so widely quarried to produce crushed rock, agricultural lime, and building stone that exposures of both fresh and weathered material are easy to find.

Limestone differs from other sediments in the way it is formed. Sandstone, conglomerate, and shale are made up of fragments, granules, or silt washed in or otherwise deposited in the place where they were consolidated into stone. These the geologist calls clastic sediments. Most limestone, however, is basically a precipitate of calcium carbonate from calcium salts dissolved in water, usually salt water, although there are freshwater limestones. The precipitate, in the form of microscopic crystals of calcite, settles on the bottom. In time it is capable of forming layers miles thick.

In this limy ooze, shells of brachiopods, clams, oysters, and snails accumulate. They, too, construct their homes of calcium carbonate stolen from the water. In the same way, cephalopods, crinoids, calcareous sponges, some algae, protozoa, blastoids, echinoids, bryozoa, and even starfish add their calcareous hard parts to the ooze. Coral reefs, made up in a like manner of the accumulated remains of the tiny coral animal, form islands miles in diameter. Such fossil coral reefs from the Silurian period are often quarried in the Midwest where they outcrop at the surface as prehistoric islands.

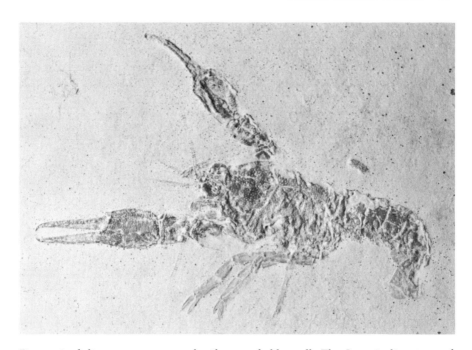

Fine-grained limestone preserves fossils remarkably well. The Jurassic limestone of Eichstatt, Germany, has faithfully preserved even the antennae of this lobster-like crustacean.

The empty shells and hard parts become incorporated in the limy mud on the bottom. Animals and plants lacking calcareous structures, such as trilobites, graptolites, and plant leaves, are buried in it, too. Calcium carbonate recrystallizes easily, so that the layer of soft lime with its embedded organisms gradually turns into limestone.

Some limestones are massive; they do not part easily into thin slabs. These are limestones that have formed over a long period of time and under constant conditions. But conditions in other prehistoric seas must have been more changeable, because in them formed thin layers of limestone interbedded with thin shales or limestones of different composition. These thin layers are often rich in fossils or almost entirely composed of them. Thick layers of Burlington limestone, quarried in Iowa and Missouri, are made up of clearly visible crinoid stem segments and plates of the crinoid cups. Scarcely any cementing material is visible in the limestone.

Shells of clams make up most of the substance of a distinctive limestone found in Florida that is used as a building stone. A limestone whose major

A drop of hydrochloric acid will fizz violently when placed on limestone. This is the best test for limestone or shale with a high percentage of calcium carbonate as cementing agent.

constituent is one type of fossil is known by the name of this fossil; examples are crinoidal limestone, blastoidal limestone, and coquinal limestone. What is commonly called coquina rock is a rock formed not necessarily of coquina clams, but of broken pieces of any shell, or of coral fragments with little visible cement and little filling of cavities between shells.

Limestone is a good final resting place for organisms. While mollusks and trilobites are often flattened and disorted in shales, they remain plump in limestone. Shale is still the place to find fossils of the soft-bodied and rare animals, but limestone is not a bad second in this respect. The Bertie limestone, through which canals were built a century ago in New York State, yielded beautifully carbonized fossils of eurypterids, complete in detail even to their legs.

The quarries of Solnhofen, Germany, have produced a fine-grained limestone for more than 400 years. It is still used for lithographic stones. On these slabs of Jurassic sediments are occasionally found extraordinary fossils, such as the feathers of birds. Fish with all scales in place are

These small nodules of limestone occasionally contain tiny silicified insects remarkably well preserved. California.

fairly common, as well as insects and shrimp with antennae intact, horseshoe crabs, and even jellyfish.

The famous crinoid locations at Le Grand, Iowa, discussed in Chapter 1, and at Gilmore City, Iowa, occur in layered limestone. Thousands of coral specimens have been collected from Devonian limestone beds in northern Iowa and at the Falls of the Ohio River near Louisville, Kentucky. Much of the state of Texas exposes a fossil-rich layer of Cretaceous chalk, which is a soft limestone. The word Cretaceous, in fact, comes from the Latin word for chalk. Another kind of limestone formed of a soft, very limy mud is called marl. It is often filled with fossils.

Many of the same types of fossils can be collected from the limestone as from shale, with a few exceptions. Graptolites are commonly found in dark shales, always compressed. They were not found in limestone until 1890, when complete, uncrushed specimens were etched from the rock. These provide good material for scientific study, but they are rare. Leaf fossils are mostly confined to shales, probably because most of them were deposited in freshwater or ocean-shore environments where limestone would not be likely to form.

Large corals are generally found in the limestone formations which they helped build. Pyritized fossils are rare in limestones but are not uncommon in shales. On the other hand, silicified fossils are rare in shales, while beautiful specimens have been collected from limestones, such as the exquisitely preserved brachiopods from the Glass Mountains of Permian age in Texas, or silicified Ordovician trilobites from West Virginia. Miocene limestone nodules found in one area of California often contain silicified insects, preserved as perfectly as the insects in Baltic amber, with every antenna and eye facet in place.

Dolomite

Dolomite is a half brother of limestone. It is a calcium magnesium carbonate in which part of the calcium of limestone has been replaced by magnesium. This replacement is believed to have taken place while the carbonate precipitate on the sea bottom was still soft. Much recrystalliza-

Poorly preserved snail typical of fossils found in dolomite. This specimen is both mold and cast; the mold of the outer shell is apparent as radiating lines; the cast of the inside of the first whorls protrudes in the rear.

tion of the stone occurred, a process which damaged or even destroyed most of the small, delicate fossils.

Casts and molds are more common in dolomite than in limestone, so much so that it is rare in dolomite to find fossils that still show original hard parts. Dolomite is usually grayer than limestone and may show crystal-lined cavities that once contained fossils. A drop of dilute hydrochloric acid in limestone will produce violent bubbling and an audible fizzing; a drop on dolomite will produce only a bubble or two unless the acid is warmed or the rock is powdered.

Calymene, the trilobite so widely collected from Milwaukee, Chicago, and Grafton, Illinois, is found as an internal cast, or steinkern, in the dolomite. Steinkerns of snails, corals, and brachiopods are associated with the trilobites, as well as poorly preserved internal casts of crinoids and cystoids. This belt of Silurian dolomites is exposed from Iowa to New York. Other Paleozoic dolomites outcrop in the Rocky Mountains.

Chert

Nodules of chert, an impure flint, are found in limestone formations. Chert is light-colored and opaque and breaks readily into sharp flakes. It is one of the materials the Indians used to make arrowheads and spearheads. It is found as fist-sized lumps scattered irregularly throughout some limestones, particularly those of Mississippian and Pennsylvanian age. After years of weathering, the nodules stand out as bumps and prickles on the strata, accumulate in the talus below the limestone bluff, or remain defiantly in the soil to annoy the farmer.

Chert must have formed when the limestone sediments were still relatively fresh and unconsolidated. The Burlington limestone of Mississippian age has several prominent chert-bearing horizons in it. The surrounding limestone is coarse-grained, and contains only occasional tough, hardy fossils such as fish teeth, crinoid cups, and heavy brachiopods. The chert nodules, when broken open or sectioned, disclose large, delicate brachiopod shells which are perfectly preserved. They extend to the edge of the chert nodule but do not continue into the limestone. The protozoa, brachiopods, and delicate, lacy bryozoa that are common in the chert are not found in the limestone, though obviously they must have been there before being dissolved away where not protected by the siliceous chert.

A chance break may expose enough of a fossil to identify it, but full realization of the beauty of chert fossils comes only from thin sections carefully cut, ground, and polished. "Rice agate," neither rice nor agate, is a brownish-gray chert loaded with protozoa found in Pennsylvanian

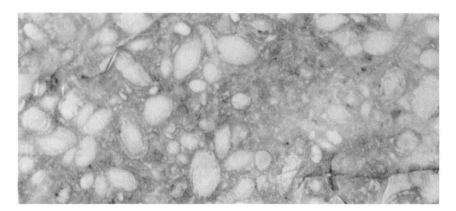

Chert may contain well-preserved fossils, such as these protozoa *Triticites*, named "rice agate" because of the shape and size of the protozoa. Pennsylvanian; Red Oak, Iowa.

Ammonites from the Cretaceous of South Dakota. These were found in concretions. The specimen in lower right still has part of the concretion attached. Lower left is a specimen showing suture lines where the mother-of-pearl layer of the shell is missing. (Photo South Dakota Department of Highways)

deposits of western Iowa and occasionally turned into cuff links or ring-stones. It polishes well, as will any fossil-bearing chert.

Concretions

Concretions are found in shale, occasionally in sandstone, and even in coal. If coal balls are described as concretions, coal may also be included. Concretions are most commonly composed of calcium carbonate or iron carbonate, although they may occur in sediments that are lacking in either carbonate. Their surface is usually curved and even spherical. They may be compared in form to a french-fried shrimp or onion ring. Such an object, dipped repeatedly in batter, builds around itself layer after layer until gradually it loses all except the rudimentary outline of its shape. Some concretions are plainly layered, while others may have grown more subtly —perhaps, like the fat lady in the circus, adding to their bulk from within. What causes concretions to form around fossils still challenges paleontologists.

Not all shales contain concretions, but those that do often contain pro-

Massive gray shale with concretion in place. The iron carbonate concretion quickly weathers out of the soft shale and darkens to a reddish color. Pennsylvanian; Terre Haute, Indiana.

Concretions may reach mammoth size, such as this one weathering free along a Pennsylvania road.

Illinois and Indiana are famous for the superb plant fossils found in concretions such as this. When struck on edge, they break along the fossil surface, exposing two halves. *Neuropteris gigantea*; Pennsylvanian; Braidwood, Illinois.

Split concretion showing the rough outside and the seed-fern leaf *Neuropteris scheuchzeri* inside. The layering of the shale in which the concretion formed is evident in this specimen. Pennsylvanian; Mazon Creek, near Morris, Illinois.

digious numbers of them. Concretions are not common in early Paleozoic rocks, but they suddenly proliferated in the Pennsylvanian period. Many shales exposed in midwestern coal mining contain them. Rocks of later periods, particularly the Permian, Triassic, and Cretaceous, are good sources of large concretions containing fossils of leaves and animals.

Concretions are easy to see and to collect from shale. They weather free until they litter the shale banks, or they wash into nearby streams. Any gray or reddish rounded rock found in shales or near shale exposures, therefore, is worth breaking open.

Strip coal mining in the Midwest has exposed thousands of acres of shale, from which millions of beautiful Pennsylvanian plant fossils have been taken. The Mazon creek area in Illinois is the best known, but similar plant fossils in almost identical concretions have been found several hundred miles south in Illinois, in western Illinois, at Terre Haute,

Not uncommonly, sandstone concretions containing large leaves are found weathering out of Cretaceous formations in Kansas. *Populites*; Ellsworth, Kansas.

Indiana, and as far away as Mineral Wells, Texas. Some 500 species of plants have been described from these concretions. "Fern fossils" in concretions have also been unearthed at Dudley, England. In the same Pennsylvanian concretions that contain the plants are occasionally found animal fossils, such as horseshoe crabs, insects, worms, fish, and, rarely, amphibians.

Other coal mines of the Midwest yield extremely large concretions which, when cracked open, often reveal a center filled with well-preserved brachiopods, snails, clams, and goniatites (small coiled cephalopods). Some of these are pyritized, particularly in western Illinois coal mines.

Among the most spectacular concretions are the septarian nodules of Knoxville, Iowa, which are related to those described above because they undoubtedly formed around a fossil. Shrinkage of the concretion caused cracks to form, radiating from the center. Solutions dissolved the fossil and deposited bright crystals of calcite along the walls of the cracks. It is from these walls, called septaria from the Latin word for partition, that the nodules get their name. Smaller nodules of this type, when sawed across to show the starlike pattern, are prized by mineral collectors.

The Cretaceous period abounded in shales and their progeny, concretions. Giant concretions weather out of shales in South Dakota, sometimes with a beautifully preserved ammonite at their center. Others contain large clams, oysters, or belemnites. The concretion may also display a septarian pattern, or it may contain golden barite crystals in the cavities.

Other marine fossils, notably ammonites, are found in large concretions in the Eagle Ford shale of the Cretaceous period exposed and quarried around Dallas, Texas. Coastal areas of California and Oregon have produced marine fossil-bearing concretions of similar age.

Fossil leaves similar to modern ones are found in concretions of Cretaceous age that weather out of sandstones and shales in a belt extending from the Dakotas down into Kansas. These are often large and unwieldy. Fossil fish, wonderfully preserved, are found in Brazilian concretions. English shales of Cretaceous age abound in ammonite-bearing concretions.

Jurassic and Triassic concretions are uncommon in the United States but elsewhere in the world carry fossils similar to the Cretaceous ones.

More recent rocks contain concretions, too. The fossil crabs of Washington State are found in cannonball-shaped concretions that lie in soft sandstone of Oligocene age. These crabs are also found as well-preserved

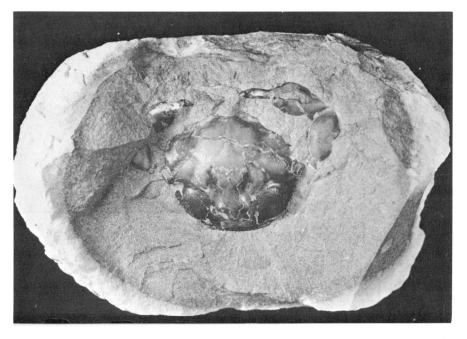

Crab fossils are found in concretions from sandstones of Oligocene age in the Pacific Northwest. *Zanthopsis vulgaris;* Washington.

fossils in the sandstone with no trace of concretion around them. Other marine fossils are common in the sandstone, but the concretions rarely contain anything but crabs. Even the recent Pleistocene has concretions: in Ontario, small fossil fish are found in concretions of that period.

COAL BALLS

Coal balls can be considered concretions, as they are rounded masses of a mineral different from the surrounding rock and deposited before consolidation of the host rock, which is coal. The compost of Coal Age forests settled in the swamps, and calcium carbonate infiltrated masses of matted vegetation, forming the coal balls. As the plant debris was compressed, these rounded masses were already petrified and remained as swellings in the coal seam.

These coal balls are rounded or lenticular, from fist size up to giants weighing a ton, but they seem to average basketball size. They are unmistakable when found embedded in the coal seam: nothing else so large and solid and round exists in the coal. The smaller ones are scooped up with the coal, separated in the washing plant, and discarded on the dump. Very large ones are left in the mine.

Coal balls occur sporadically. One Kansas coal mine ran into so many

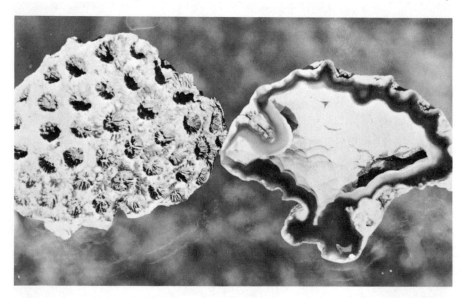

This chalcedony-lined geode was once a coral. Covered with clay and soft limestone, the coral dissolved, leaving only a mold of the outer surface. Later, silica-rich waters deposited layers of agate and chalcedony in the cavity. This specimen has been cut in half, showing exterior and interior. Miocene; Tampa, Florida.

that it became uneconomical to rid the coal of them, and the mine was closed. But in a nearby mine there were none. Mines in Iowa, Kansas, Illinois, Indiana, and Kentucky are particularly rich in these concretions. The Mazon Creek region in Illinois, famous for its fern-fossil concretions, is currently producing a number of coal balls. They are partly pyritized. Many mines produce such pyritized coal balls; they are useless for research as they cannot be properly "peeled" or sectioned.

Coal balls can be appreciated only when they are examined under a microscope. Their value lies in the perfectly preserved cell-for-cell petrifaction of the original woody tissues. The cell walls are still there, as are spores still in the spore sacs of 275 million-year-old fruiting bodies. Preparation of these fossils is treated in Chapter X.

Geodes

Geodes are not sources of fossils, but some were originally fossils. A geode is a nodule of stone having a cavity lined with crystals or minerals.

Once fossils, now geodes with hollow, quartz-lined centers. In becoming geodized, these unusual fossils grew greatly in size. This geodized crinoid stem was once the size of the large crinoid stem segment in the foreground. Mississippian; Brown County, Indiana.

It originated as a hole: possibly a hole left in rock by a dissolved fossil, or it may have been the hole inside an existing fossil. Water rich in silica percolated through the rock, depositing layers of silica around the edges of the cavity. Sometimes, this process continued until the cavity was filled with layers of agate or crystalline quartz. Often it stopped before this stage was reached, and a crystal-lined, hollow, quartz "geode" was left in the cavity.

In a few areas of Indiana and Kentucky, Mississippian shales release geodes that are unmistakably of fossil origin, for they are shaped like high-spired snails, crinoid cups, brachiopods, and corals. However, they are giants. Crinoid stems may be as big as a man's wrist and brachiopods the size of oranges. The exterior is rough and cracked. It appears that the fossils had swollen to five or ten times their normal size. This must have happened while the silica was forming—perhaps it formed on the outside of the organism, pushing away the surrounding shale as it grew. The center was then filled with quartz crystals. No actual shelly material remains, only the obese casts of the fossils.

V

TIME BEFORE TIME

Some words used in this book, such as "Paleozoic" and "Mississippian," label fossils by one of their most important characteristics—their place in the span of time. When a fossil played its part in the parade of life is just as significant in geological history as when certain men and nations played their parts in human history. Geologic time, far longer than historic time, reaches back into the past several billion years, and it is also, of course, less definitely documented than written history.

Geologists work with two kinds of time: relative time and absolute time. The difference in the two forms is simple. Relative time places organisms or earth events in a time sequence, in an orderly series, just as we would call the roll of the rulers of England, saying that James I came after Queen Elizabeth, and Charles I after James I, etc. The Bible chronicles events in history in the same relative fashion, not by giving dates but by placing men and events in chronological order.

Absolute time, on the other hand, reckons elapsed time since a rock or fossil was formed by measuring the rate of decay of radioactive elements and expressing the change in terms of years.

GEOLOGIC TIME CHART

ERA	YEARS BEFORE PRESENT		Important Life Forms	Succession of Life
CENOZOIC	RECENT	11 THOUSAND		Mammals spread with development of modern grasses. Mollusks important on land and in the seas
	PLEISTOCENE	3 MILLION		
	PLIOCENE	13 MILLION		
	MIOCENE	25 MILLION		
	OLIGOCENE	36 MILLION		
	EOCENE	58 MILLION		
	PALEOCENE	63 MILLION		
MESOZOIC	CRETACEOUS 135 MILLION			Decline of dinosaurs and ammonites Development of modern invertebrates
	JURASSIC 180 MILLION			Earliest birds Reptiles abundant Cephalopods
	TRIASSIC 230 MILLION			Comeback of ammonites
PALEOZOIC	PERMIAN 280 MILLION			Decline of ammonoids Extinction of trilobites
				Plants, insects, and marine invertebrates
	PENNSYLVANIAN 310 MILLION			Crinoids, blastoids, and brachiopods important
	MISSISSIPPIAN 345 MILLION			Trilobites waning Brachiopods, corals, first land plants Fishes
	DEVONIAN 405 MILLION			Brachiopods, corals, crinoids trilobites
	SILURIAN 425 MILLION			Trilobites, corals, graptolites, crinoids, and nautiloids important
	ORDOVICIAN 500 MILLION			
	CAMBRIAN 600 MILLION			Trilobites, inarticulate brachiopods, and gastropods important
PROTEROZOIC and ARCHEOZOIC ERAS $4\frac{1}{2}$ billion years				

RELATIVE TIME

Fossils and their associated rocks provide much of the data by which relative time, or the relative sequence of events, is calculated. Study of the evolution and interrelationships of fossils and of the nature, position, continuity, similarity, and alteration of the rocks, interpreted in the light of the assumption that conditions in the past were little different from those of today, is the basis of the system of eras, periods, epochs, and ages which is the framework of relative geologic time.

Relative geologic time is divided into five eras, beginning with the Archeozoic ("beginning life") and the Proterozoic ("primitive life") eras. These two eras occupy nearly 4 billion of the $4\frac{1}{2}$ billion years of the estimated existence of the earth. Their fossils are few, faint, and often problematical; they do not interest the amateur. But out of this dimly understood void, life sprang abundantly with the dawn of the next era, the Paleozoic ("ancient life") era. Like subsequent eras, the Paleozoic is divided into shorter units, the periods, which themselves are divided into epochs and ages. The first Paleozoic period is the Cambrian, and in fossil literature the term "Pre-Cambrian" is often used to designate time before the Paleozoic.

The Paleozoic era was followed by the Mesozoic ("middle life") era, often referred to as the Age of Reptiles, then by the Cenozoic ("recent life") era, the Age of Man, in which we are living. The Cenozoic is divided into the Tertiary and Quaternary periods, but is more commonly described by reference to epochs, which are subdivisions of the periods.

The Grand Canyon of the Colorado river is a monument to relative time. Here deposition built a remarkably complete series of strata lying on ancient metamorphic schists. As earth forces slowly raised the plateau formed by these strata, a dauntless river, fighting for its existence, cut through the noble pile of rocks, exposing, like the sheets of a desk calendar, the strata that record by their thickness, composition, and fossils the geological history of this region for the last 2 billion years.

Some strata are missing altogether or appear at only one or a few places in the Canyon. But these parts of the whole history can be tied into one chronology by tracing a few key formations, such as the Redwall limestone, and with it as a fixed point determining what is lacking. This is the way a historian works, gathering his documents where he can, fitting them into place, and then drawing conclusions. Similarly, the geologist studies the composition and texture of the stone, studies the comparative

The major divisions of geologic time, their duration, and some of the major life forms of each era.

development and character of the fossils in it, and perhaps supplies a missing fact from study of comparable formations elsewhere that may be more complete in some detail. From these data he arrives at a place and a time for the formation and its fossils in the larger context of facts, marshaled by all geological and paleontological research.

The Grand Canyon is a magnificent record of such events in time because of the long vertical exposure of strata going back into the Pre-Cambrian even though the Mesozoic record must be supplied from nearby formations because it has been eroded from the path of the Canyon. Furthermore, even the Pre-Cambrian Grand Canyon system of sedimentary rocks resting on the metamorphosed schists carries algae fossils, and all the rocks above it contain fossils, which makes the work of correlating them much easier. Certain fossils characteristic of a widely distributed rock formation and present only in that formation are used to identify that formation wherever it may appear. These are called *index fossils*. Their presence in a formation helps geologists to tie together the evidence of identity of formations even though they may be broken or disguised by erosion, by geological faults, or by differing conditions.

One of the most useful tools used by the geologist to date formations is *correlation*, which is the placing together of the relationships of unlike rocks even though they may have been formed at the same time and in adjacent areas. An ocean, a shore, and a brackish swamp may have been geographical neighbors at some time far in the past, just as they are today. Yet each would spawn its own particular fossils because conditions in each made life possible for specific organisms.

Facies

Geologists have a name for a part of a rock body that has such a relationship to other parts of a rock body. They call it a *facies*, and define a facies as the general appearance or nature of the one part as compared with the other parts.

A series of modern-day facies could be created by instantly petrifying a part of Florida and entombing its denizens in sediments hardened into an unbroken sheet of rock of the same age. Part of the rock would represent a freshwater facies that might include fossils of freshwater fish, clams, crayfish, and plants in a dark shale that was once mud from a river that fed into a lake. Another nearby facies in the continuous rock layer would include tidal-pool facies fossils such as starfish, tiny saltwater fish, sea anemones, oysters, barnacles, and crabs. Several hundred yards away a new facies might appear, the deep-water one, marked by certain shellfish, corals, larger fish, jellyfish, shark teeth, etc. Each facies, although all are contemporaneous, has a distinct fauna.

Comparable fossil facies exist, such as those in the Silurian coral reefs that are exposed over a wide area from New York to Iowa. A quarry wall where such a reef is being worked may show the top of the ancient coral island with its included broken fossils of mollusks and some arthropods. Farther along in the quarry would be a facies of large, broken chunks of corals, brachiopods, and tough-shelled clams from the edge of the reef, where the surf once pounded. Next to it would be a third facies of the deep water beyond the reef, perhaps a solid limestone with crinoids, bryozoans, and trilobites.

Geologists have their own words to indicate the relationships of rocks and the relative age of the rocks. The amateur will find reference in the professional literature to rocks, for example, as being from the Lower, Middle, or Upper formations of a period, but the period will be divided chronologically into early, middle, and late. The Lower Cambrian rocks will be from the early Cambrian, and will be the oldest of that period. Rocks of a period are called a *system* and those of an epoch, the next smaller unit, are a *series*.

Formations

But in practice and in principle, it is common and necessary to identify rocks as a formation, the name of which is usually a compound of a typical location and the rock type, such as "the St. Peter sandstone" or "the Trenton limestone." If the formation consists of more than one type of rock, for instance, if it is of shale and limestone, it may be called by a more general name, such as "the Supai formation," which appears in the walls of the Grand Canyon, or "the Morrison formation," famous for its dinosaur fossils.

Even though in no one place do all rocks of all ages appear as a complete chronicle of the past, in many ways geologists have questioned the mute evidence of the rocks and have arranged it into a coherent story that is concerned more with deeds than with exact dates. Time is reckoned as incidental to the broad picture of earth history. But occasionally geologists are surprised with an incredibly detailed story written in the rocks. Mecca, not the Moslem holy city in Arabia, but a hamlet in western Indiana near the Illinois border, stands where a swampy forest grew 300 million years ago. Close by its forest of horsetails, tree ferns, and scale trees lay an estuary of an inland sea that then covered much of Illinois and parts of Indiana.

Today that ancient Pennsylvanian swamp is known only from a deposit of black shale, but in that deposit scientists from the Field Museum read the story of four years—four years far back in the hundreds of millions that have passed since that time.

Some force, perhaps an earthquake, allowed great quantities of salt water to flood into the swamp, felling the trees. Sharks, primitive fish, and shelled creatures wandered into this new environment, probably attracted by the food washed loose. They flourished until a dry season came. Then the pools shrank, the trapped creatures gasped, fought, mutilated each other, died, and were buried under the thick mat of floating vegetation that turned to black mud and then to shale. The shales show that this happened four times, four seasons of life in the rainy season, death in the dry season. Then the coast settled, the waters grew permanently deeper, and brachiopods, cephalopods, and other invertebrates left their quiet fossil record in the lighter-colored shale that tops the black shale of the death pools. Museum experts under the direction of Dr. Rainer Zangerl, chief curator of geology in the museum, excavated this ancient pond and brought back the remains of huge sharks and thousands of mutilated fish and other fossil creatures from Mecca.

ABSOLUTE TIME

Opportunities to come as close to a geological event as that at Mecca are few. Certainly it is generally far beyond the scope of the system of relative dating described above, which is one of the two measures that geologists have when they speak of time. The other measure they use is a clock that records absolute time by measurement of changes in radioactive elements in the rocks. Such elements decay at a constant rate, the works of an atomic clock that ticks steadily but so slowly that its face is marked in thousands and millions of years. This great-great-grandfather of all clocks is capable of timing the span of life from the amoeba to man.

Radioactive decay or nuclear fission, the force that powers the atomic bomb and nuclear power plants, causes transmutation of a radioactive element into a new element. Because this decay takes place at a predictable rate regardless of heat, pressure, and solution, measurement of the amount of the parent radioactive element and the transmuted daughter element in the rock makes it possible to set a quite exact age in years for the rock.

Fission changes uranium into lead and helium, potassium into calcium and the rare gas argon, and rubidium, an element chemically close to sodium, into strontium, the element that puts brilliant reds into fireworks.

In the same way, one radioactive form of carbon turns into a more stable form of the same element.

The Atomic Clock

Age determination by nuclear methods has been well described by Dr. Dr. Edward J. Olson of the Field Museum in the following quotation, which is used by permission. He wrote in the museum's Bulletin:

Suppose we had a large box with 6,400 green marbles in it. Then imagine that by some process in exactly one year half of the marbles had turned red. This leaves 3,200 green ones and 3,200 red ones. Suppose that in one year half of the remaining green ones become red, leaving 1,600 green and a total of 4,800 red. If the process continues in this manner we may then construct a table:

Passage of Time	Green	Red	Red Divided by Green
0 years	6,400	0	0
1 year	3,200	3,200	1
2 years	1,600	4,800	3
3 years	800	5,600	7
4 years	400	6,000	15
5 years	200	6,200	31
6 years	100	6,300	63
7 years	50	6,350	127
	And so on		

If we know that this process goes on with regular precision, we could look at such a box, count the reds and the greens and then say how long the marbles had been sitting there. For example, if we found 6,200 red ones and 200 green ones we could say that the process had been going on for five years. In fact, we need not necessarily go through the trouble of counting all the marbles. The right-hand column in the table show the quotient of reds divided by greens. Thus, we need only take out a random sample of a few hundred marbles and count the reds and greens, divide the former by the latter and, if our sample is average, we should obtain a value close to 31—a time of five years. This process goes on until the last green marble has shifted to a red color. At that time the clock may be considered to have run down. What we have just described, in a fairly simplified form, is the so-called atomic clock upon which the much publicized methods of radioactive dating are based.

Rather than by marbles changing color, the actual atomic clock oper-
ates by atoms changing to other atoms. The time required for half the
population of atoms of one kind to change to another kind is called the
half-life.

Before going on let's look once more at the box of marbles to clear
up another definition. Let us imagine that every time a green marble
converts to a red one it gives off a loud clicking sound. During the first
year we would observe 3,200 clicks, or an average of around 62 per
week. This is moderately noisy. During the second year, however, there
would be only 1,600 with 31 per week on the average. During the third
year there would be only 800 clicks, or about 15 per week; and so on.
Thus the rate of noise-making would drop off year by year until it
finally stopped. At any time during the life of this clock we would have
a definite noise level. This we call the level of activity. In the case of
atoms this is called the level of radioactivity. So far then we have two
methods to measure time. We might, as mentioned before, count a
sample of red and green marbles and figure the time from that; or we
might simply count the number of clicks per week, or per day, etc., and
figure the time from the rate at which they are being produced. In the
first method we need not necessarily know how many green marbles
were present in the beginning since we are only measuring the quotient
of reds divided by greens, which will be the same no matter how many
greens were there originally (if you don't believe me you might give
it a try, starting with, say, 10,000 green ones). We need to know only
the half-life, which in this example is one year. In the second method,
however, we have to know the original population of greens in order
to correlate the level of activity with the age of the system. If the half-
life is only a year, or an hour, or, as in the case of some atoms, only a
few seconds, it is obvious that such clocks will "run down" in a short
time and be of little value. To use such weak-springed clocks we have
to have an extremely delicate chemical method to analyze exactly the
number of green atoms and red atoms. Once the number of green ones
has fallen below our ability to separate them in the laboratory, the
clock is, for all purposes, dead even though there might be some few
green atoms still present. The same is true if our ability to detect
the clicks per unit of time is limited by our laboratory devices.

Minerals from igneous rocks are generally the most satisfactory for
age determination, although an earthy mineral, glauconite, associated with
sedimentary rocks, is also used. The specimen to be tested should be a
single mineral, unaltered since it was formed, and it should contain a
measurable amount of both the parent and daughter elements.

Potassium changes into argon, a gas, with a half life of 1.32 billion years, which makes measurements of these two elements a suitable means of dating rocks as old as 4.5 billion years, the estimated age of the oldest material in the crust of the earth. In this test, typical of several frequently used, the rock sample, mica or a feldspar, is fused to free the argon gas, and the amount of gas is measured by a mass spectrometer, which by magnetic means isolates the element to be measured. Other methods measure the ratio of rubidium and strontium, which have a similar relationship, or of uranium and thorium.

The Carbon-14 Method

For organic objects such as wood less than 40,000 years old, measurement is made of a certain form of carbon, Carbon-14. This radioactive form of carbon is the product of the action of cosmic rays in the upper air. It unites with oxygen to form radioactive carbon dioxide, which is taken up by living plants and then by animals that eat the plants. It is found in a constant proportion in all living tissues, decay of the radioactive carbon being balanced by constant replacement of cells. The intake ceases when the organism dies, but the decay of its radioactive carbon continues. After 5,565 $\pm$ 30 years (the sign $\pm$ means plus or minus that amount for experimental error) half of the carbon will have changed to the stable form. Measurement of Carbon 14, therefore, makes possible precise age determinations on some relatively recent fossils and carbon-containing materials, including carbonate rocks. Archaeologists carefully remove hitherto "worthless" charcoal or wood bits found along with pots and arrowheads for C_{14} dating of their finds.

Carbon-14 dating of the wood of logs uprooted by the most recent continental glacier in Wisconsin established that the event happened 11,400 $\pm$ 200 years ago. Until this test set a more definite figure, it had been assumed that the glacier made its call 25,000 years ago.

Many measurements have been made by these methods and have been used to confirm and make more precise the findings of paleontologists all over the world. Perhaps as good an example as any, which seems more personal because it casts light on the genealogy of man, comes from the discoveries made by Dr. L. S. B. Leakey in Africa.

When Leakey discovered fossils of precursors of man in Olduvai Gorge in Tanzania, the age of the rocks in which they lay became a matter of major paleontological importance. Luckily, the fossils were preserved in volcanic ash, so that fresh igneous minerals were present right with the fossils.

Potassium-argon tests set the age of the volcanic ash at 1.75 million years. This determination was cross-checked by another method that made use of fragments of volcanic glass in the rocks. These were polished and examined under the microscope for tracks only a few atoms wide made in the glass by fission of uranium atoms. After the tracks had been counted, the samples were irradiated to cause all their contained uranium to produce tracks. These were counted, and comparison of the two sets of numbers gave a measure of the extent of uranium decay. From it the age of the rocks was estimated at 2 million ± 30,000 years. After allowing for innate sources of error in the two methods, the experimenters regarded their results as substantial confirmation of an age for the fossils of nearly 2 million years.

Some ingenious fellow has reduced the incredible dimensions of geologic time to more familiar terms by bringing it within the scale of a calendar year. If the history of the earth were compressed into twelve months, the first eight months would be represented by virtually blank pages in the calendar. These would be the Proterozoic and Archeozoic eras. September and October would see the development of algae and bacteria and the gradual appearance of the invertebrates as fossils. Mammals would not come on the scene until mid-December. Man himself would not have evolved until the last few minutes of the last hour of the last day of December in this hypothetical year, and what we know as written history would span little more than the last minute of the year.

Such is geologic time; such is the theater of paleontology.

VI

WHERE TO LOOK
FOR FOSSILS

Much of the United States is underlain by sedimentary rock which contains fossils, but the northern third of the nation is blanketed with glacial debris, sometimes as much as one-hundred feet thick, and even the unglaciated regions wear a mantle of soil. For this reason, the fossil collector will have to seek places where the rock is exposed by excavations of one sort or another.

These include quarries, coal and metal mines, roadcuts and railroad cuts, foundation excavations, tunnels, and canals, as well as such natural exposures as river banks, beaches, and mountain slopes. For certain special types of fossils the collector will turn to bogs, tar pits, seeps, and the like. This chapter will discuss each type of site, with suggestions about methods of collecting and what may be expected in each.

LIMESTONE QUARRIES

A quarry is an open pit from which rock is removed for use in road building, construction, and concrete work, and for agricultural lime, stone slabs, and the manufacture of cement.

Vertical or steeply tilted rock layers are sometimes found in quarries, especially in mountainous regions. Kentland, Indiana.

The Geological Surveys of most states will supply a list of producing quarries in their areas. Large-scale topographic maps show such excavations, both working and abandoned quarries, but such maps are often old. Local construction and paving companies can supply information, as they depend on nearby quarries for material. Most quarries have a tall crusher plant, surrounded by piles of rock, which is usually easy to find.

A typical quarry may be several square blocks in area and have a working face from thirty to well over one-hundred feet high. Its depth is dependent on the thickness of the usable layer of stone. The limestone which is quarried is usually in thick layers and not very fossiliferous.

The rock may look the same from the top to the bottom of the quarry, but often two or three important formations are included in one quarry wall. One formation only two feet thick may carry numerous fine fossils, while the rest of the rock is barren. With patience, these layers can be recognized in the quarry wall, and then in the broken rock on the quarry floor after blasting.

Generally, collecting from the freshly blasted rock is not very profitable unless the quarry is in dolomite rather than limestone and most of the fossils are casts and molds, which must be broken from fresh, un-

Freshly blasted limestone in a quarry will seldom be a good site for collecting. No weathering has occurred to expose fossils, and the rock is usually coated with dust or mud.

weathered material. Molds will break in half, and both halves should be saved to make fossil casts with plaster, rubber, or plastic at home.

The mold itself will usually reproduce the outer shape of the fossil. But it may contain an internal cast, a steinkern, which can properly be identified only if it has been found in the cavity of the external mold. It will often bear little resemblance to the outer appearance shown in the books. This is especially true of brachiopods. The relationship of a steinkern and its parent mold should be made clear in labeling.

Some types of fossils will not stand weathering, so whenever they are found in limestone, it must be in fresh material. These include carbonized films of crustacea or worms, or such delicate specimens as crinoid crowns.

Some fossils pop loose from freshly broken rock and can be collected by breaking up the rock with a sledgehammer. The Mississippian crinoid calyxes found in Iowa, Illinois, and Missouri are collected this way. Some brachiopods also break out cleanly.

The best collecting in the quarry is from weathered material. Five years of weathering on what once appeared to be an unfossiliferous block of limestone can magically produce a coating of fossils. Furthermore, softening of the limestone makes removal of these fossils much easier. If the limestone weathers even longer, the fossils may drop out and can be collected on the ground. Too much weathering will obliterate fossils, but new ones are constantly exposed.

Weathered limestone can be found in several spots. Occasionally, elephant-sized boulders of solid rock are blasted out. These are too large for the crusher and not worth blasting to a usable size, so they are bull-dozed into an unused area of the quarry floor. There they weather away, waiting for a fossil collector. Older areas of the quarry will have a certain amount of debris on the floor, including pieces that have scaled from the quarry wall. Collecting along a wall is always a risky business; a fossil from above may collect you on the way down. (See Chapter 7 for more about safety precautions on collecting trips.)

At times new roads are built into the pit, requiring fill, and artificial hills are built on one side of crushers so that the trucks can dump directly into them. These roads and hills are made of the native stone, and years of weathering may create fine collecting along their side slopes.

To quarry the merchantable rock the quarry operator must remove

A museum team collecting Devonian fish and scorpions at a small quarry in Montana operated by the Field Museum.

overlying dirt and debris. He may also have to remove some layers of shale or impure weathered limestone; some quarries have mountains of this waste material. It may be the most fossiliferous, and it weathers the most rapidly.

Gray, muddy-appearing piles are shales, and yellowish or dark-gray rocky piles are limestones. All are worth examining for fossils. Good examples of such dumps are found at Waldron, Indiana, and Pegram, Tennessee. In both quarries, a thick limestone lies under the thin but highly fossiliferous Waldron shale of Silurian age. The shale that is scraped off and rejected is the real treasure trove. The barren limestone is worthless—except commercially.

Another place to look is below the bottom of the quarry. Water constantly seeps into any excavation and unless pumped out would flood the quarry. It does not take many months or years to turn a quarry into a good place to hunt ducks or catch fish but not to get road material. To keep the quarry dry enough to work, a deep pool or sump is needed. This is dug in a low spot in the quarry floor, usually a soft shale or unmerchantable limestone underlying the quarry's productive strata. But what is unprofitable for the quarryman may be very rewarding for the fossil collector who carefully examines the dumps around the edge of the sump pit.

But the piles of crushed rock that are the quarry's stock in trade hold little of value for the collector. Fossils of any size have been disfigured by the crusher's jaws until only microfossils are left.

If at all possible, when collecting from dumps in quarries, first examine the working face to determine what layer produced the fossils you are collecting. This may be important in trying to date and identify the fossils later. The quarry operator is rarely able to tell what formations or even what age rock is being quarried. But some State Geological Surveys and universities have yearly field trips. If you can obtain a field trip guidebook that includes the quarry, there may be a picture or drawing of the quarry face with the formations listed. Characteristic fossils from these layers are listed by name or even illustrated in field trip guides. Annually the Bulletin of the American Association of Petroleum Geologists list guidebooks of this kind published during the year.

SHALE QUARRIES

Shale is sometimes quarried for the manufacture of bricks and ceramics. Some is mixed with ground limestone in making cement. For some reason, shale quarries are usually called clay pits or brick pits. They are so designated on geologic maps.

Such pits can be located by inquiring of the Geological Surveys or by asking questions locally. They are rarely as deep as limestone quarries and are short-lived, designed to remove a thin layer of profitable clay near the surface before being abandoned. Booklets from the state geological survey list clay and shale producers of the state and may list the age and formation of the shale being removed.

Some fossils are best collected from fresh shale, and others are at their prime after long weathering. Puryear, Tennessee, is a classic area for collecting Eocene leaf fossils, but only by splitting extremely fresh shale. A few days of weathering and rain will destroy a leaf that has been in the rock for millions of years. Fragile carbon imprints of ferns are not uncommon in midwestern and eastern strip coal mines but disappear after the slightest weathering. Tougher creatures such as brachiopods, corals, and crinoid stems, on the other hand, will be cleaned free of matrix by weathering.

The recomended method is to find a fresh block of shale near the working face of the shale pit. Split it along bedding planes and look for soft-bodied fossils, then search the weathered piles along old pit walls or scattered around the area.

Shale pit (right center) of Rockford Brick & Tile Company, Rockford, Iowa, a notable collecting site. (Photo courtesy Rockford Brick & Tile Company)

As in limestone quarrying, unsalable shale will be hauled aside and left in heaps for fossil hunters. Thin limestone layers may occur in the shale. Frequently these are very fossiliferous while the accompanying limestone layers are not. Usually inquiry at the quarry office while registering there will disclose where piles of the shale may be found.

Many clay deposits of Pleistocene age, particularly those in the northern United States, formed at the bottom of small frigid lakes near the terminus of a glacier. They consist of finely ground rock dust produced by the glacier. Few living things braved these arctic waters, either for living or swimming, and such clays are only sparsely fossiliferous. In contrast, clay quarries in Cenozoic, Mesozoic, or Paleozoic rocks are remarkably fossiliferous. Ammonites are found in Cretaceous shale pits of Dallas, Texas, brachiopods of Devonian and Mississippian age in Indiana and Iowa.

It is profitable to check the working face of the shale pit carefully, layer by layer, At the Medusa quarry at Sylvania, Ohio, for example, a thin, two-inch stratum of Devonian shale produces most of the rolled-up trilobites in that area. Another thin layer a few feet above the trilobite horizon yields most of the complete crinoids found in the quarry, while a third layer is the final resting place of rare phyllocarids. Although the product of the Medusa quarry is basically limestone, a thick layer of shale laying above the limestone is removed separately. Trilobites and crinoids can be collected from the shale dumps, but they are mixed with barren material. If the horizon of the fossils can be located in the wall of the quarry, it can be followed, and usually the shale is soft enough to allow some digging into this layer.

Many shale quarries are worked down layer by layer. If the collector is on hand when the proper layer is exposed it can be a bonanza. A shale pit near Annapolis, Indiana, was worked in this way, and when one layer was almost completely removed, a paleontologist from the Field Museum in Chicago found that that particular layer was crowded with rare Pennsylvanian-period fish fossils. Thousands had been destroyed. While digging a ship canal on the south side of Chicago, workmen dumped a thin seam of shale along a quarter-mile of bank. Many years later a collector found that this shale contained 400-million-year-old worms preserved faithfully as a carbon film—the only ones of their kind in the world.

Most commercial shales and clays are light-gray or tan. These generally will contain marine fossils. Black shales found in some clay pits and strip mines are typical of near-shore deposits rich in organic material. They may contain fish or plant fossils. Small concretions that weather free from light-colored shales are worth cracking open; a small fossil, perhaps pyritized, may lie at the center. Clay pits operating in Cretaceous gray shales, such as around Dallas, Texas, contain huge amonite-bearing concretions that are neatly piled up in the pit, awaiting the collector.

When collecting from weathered dumps, the collector can get a clue to the layer from which the shale came by the color of the weathered material. A group of collectors examining shale piles in a southern-Indiana quarry found a few perfect blastoids weathered free. One experienced collector noticed that the large, perfect specimens were always lying on piles a shade lighter than the other weathered shales. He searched for these light spots, and found dozens of the blastoids, much to everyone's amazement.

COAL MINES

Among places where man has broken the earth for useful materials only quarries exceed coal mines in number and extent. The Midwest is particularly rich in bituminous coal mines, from central Illinois southeast into Indiana and Kentucky, and from mid-Iowa through western Missouri, eastern Kansas and on into Oklahoma and Texas. Another broad belt of fields rises in western Pennsylvania and spreads across West Virginia into eastern Kentucky and Tennessee down into Alabama. Colorado and Utah possess smaller fields. Northward from Colorado lie fields of subbituminous coal and lignite.

Coal deposits formed on the borders of the basins of inland seas, where brackish swamps with abundant vegetation existed. Where the waters were deep, limestone was deposited, often in thin strata crowded with fossils of brachiopods, clams, gastropods, crinoids, and cephalopods. As the seas drained away, the jungle flourished and then died, leaving thick masses of organic material that later became coal.

Coal mines are either shaft or strip mines. Coal seams no more than a hundred feet below the surface can be profitably mined by stripping away the overlying soil and stone to expose the coal seams at the bottom of a great pit. The coal is scooped into trucks or railroad cars and hauled away. Such mining is less costly and less dangerous than underground shaft mining, where there is always the peril of poisonous fumes or explosive dust. There is a further advantage in that all the coal can be recovered; in shaft mines, coal must be left in the form of pillars to hold up the roof.

Strip mines are usually spread over a huge area ripped into rows of herringbone hills and small ponds. In time, such a region loses its raw, gashed hideousness; trees cover the hills, and the small lakes become fishing and recreation assets. Old strip mines are designated as such on

Complete crinoid crowns, showing stem, calyx, and arms, are prized specimens. This is *Rhodocrinites*; Gilmore City, Iowa.

topographic maps and are concentrated along the swampy shoreline of the ancient basin. The largest coal field, the Illinois basin, has been strip-mined along the Indiana-Illinois border from Danville, Illinois, to Kentucky, across southern Illinois and then northward along the western Illinois line to a point opposite to Danville, as well as east and west along the Illinois river through La Salle and Fulton counties. The central part of the basin is mined by shafts to reach deeply buried coal.

Southern Iowa from the middle of the state to the western border also has many strip mines, and there are some in nearby states, as well as in southern Michigan, Pennsylvania, and other Appalachian states.

An operating strip mine will have a plant office where permission should be obtained, even to enter abandoned workings. The collector may be asked to sign a release form to relieve the operator of responsibility for injury. Strip mines present certain dangers. Vertical walls rise steeply,

Coal strip mine in action. Big dragline scoops up the gray shale overlying the coal and piles it to one side. Dugger, Indiana.

After weathering, the concretions litter the surface of the strip mine dump. Many will contain fossil ferns. Near Terre Haute, Indiana.

and the edge may have been shattered by blasting. The weight of a few persons near the edge may start a rock slide. The weathered shale becomes slippery in wet weather, and an unlucky collector may slide into the deep, cold water of a pit if he is not careful.

PITS

The working face of the strip mine is a good place to start looking, as the various strata can be examined there. If any prove fossiliferous, like rock can be searched for later in the dump, where it is easier to break. It is also fruitful to ask permission to collect from newly exposed coal seams. Fossils may be locked in the coal or right on its surface.

Petrified, often pyritized, logs and coal balls are found in the coal seams. The Coal Age trees did not have a well-defined woody structure like modern trees and show no growth rings. Invariably the logs are crushed and flattened, suggesting not only soft wood but great pressures. While the wood will cut and polish, there is little pattern. Huge wood sections found around Knoxville, Iowa, are cut and polished to show secondary mineralization of calcite that created white, fan-shaped patterns in the black calcified wood.

Logs or coal balls replaced by pyrite will soon disintegrate and are best left behind. Mines in Indiana and Illinois are the best places to look for coal balls, as are a few mines in Kansas and Kentucky. The wood is more common, especially in central Illinois, southern Iowa, and Pennsylvania.

A few coal seams may contain pyritized gastropods, cephalopods, and brachiopods. Near Farmington, Illinois, the miners were amazed by the week-end armies of collectors who came to brush off every square inch of the newly exposed coal seam with brooms. Embedded half in the coal and half in a black shale resting on the coal were fat gastropods (*Shansiella*), and occasional small brachiopods, straight cephalopods, and, rarely, large coiled cephalopods. The fossils were found in small depressions in the surface, suggesting that the organisms were washed in when sea water flooded the coal swamp. Iron and sulfur, prevalent in areas of decaying vegetation, provided the chemicals for replacement of the fossils by pyrite or marcasite.

Black organic shale occurs in some mines in a thin layer right above the coal. It will usually contain fossils of animals able to survive in shallow, brackish water, such as some fish, scallops like *Dunbarella*, brachiopods (*Lingula, Orbiculoides*), and cephalopods. Plant fossils are uncommon in black shale. The black shale is usually finely layered and easily split. Fossils found in it are often pyritized, and unless the shale

Shansiella carbonaria, the golden snail, its substance replaced by pyrite. From a Pennsylvanian coal seam at Farmington, Illinois.

breaks cleanly around the fossil they are difficult to prepare. Such pyritized fossils tend to disintegrate even after cleaning and spraying.

Fish fossils are perhaps the most sought-after fossils in the black shale. They are almost never found in any other rock layer in coal mines. Since the black shale is quickly weathered in the dumps, fossils must be collected from the freshly exposed shale and coal seams. Besides the thousands of fish, including some portions of giant sharks, collected from the black shales near Mecca, Indiana, others have been found near Braidwood, Illinois, and Des Moines, Iowa. Undoubtedly coal mines elsewhere would provide fish fossils if a thorough search was made in the black shales. These shales may appear for a few weeks during mining and disappear, only to reappear a mile away during later stripping.

WASHING PLANTS

Coal is usually trucked from the mine to a washing plant, where petrified logs, coal balls, and stray pieces of shale are removed. The waste rock is piled nearby; some poor-grade coal is often discarded, too. The entire pile is usually found smoldering through spontaneous combustion from heat released by decay of pyrite. If the fumes are not too bad, the collector can search the piles for fossils.

WASTE PILES

The biggest collecting area at a strip mine is the badlands created by dumping of overburden. All rock layers seen in the working face can be found somewhere in these dumps.

Gray shales are the commonest rocks here. They may contain plant fossils. Magnificent fronds of ferns preserved as black carbon films on the gray shale occur in the layers lying right above the coal. Such fossils must be collected by splitting blocks of the freshly mined material, because these gray shales collapse into sticky clay after a few rains. Shales in Pennsylvania and other eastern mines are a bit more durable, but the plant fossils are not as sharply preserved as the midwestern ones.

Weathered waste piles from strip coal mines are often fine collecting sites for fossil plants and marine fossils. Braidwood, Illinois.

If one block containing fossils is found, there are more around; these plant-bearing gray shales are more persistent in extent than the black shales.

After a few months or years of weathering, the hills become smooth piles of clay. The weathering, however, has released any hard fossils preserved in the shales and left them on the slopes while the clay washed away. Typical Pennsylvanian marine fossils can be found weathered free in some areas, particularly western Illinois, southern Indiana, and a few places in southern Illinois. These fossils are usually calcified, but are often badly crushed and distorted. The tiny ditches and ravines cut in the older hillsides are the best place to look, even in mining areas overgrown with grass and trees. There are always a few open ditches, even in the oldest stripped areas.

Nearly 500 species of plants have been found in the fossil-bearing concretions of the strip mining area around Braidwood, Illinois. The concretions bear the fossil in the center surrounded by shale hardened by iron carbonates in concentric layers around the fossil nucleus. Their origin, like that of the coal balls, is mysterious and the cause of much debate. Iron carbonate is highly resistant to weathering, and as the soft gray shale washes away, the hard concretions remain behind. When "ripe," the concretions conveniently turn a warm, red-brown color to contrast nicely with the remaining gray mud.

Near Braidwood, Illinois, is the mine whose concretions have provided science with more than a hundred new species of soft-bodied animals and insects, in addition to plants. Large concretions are found weathered loose on the waste piles of strip mines in western Illinois. Each may contain hundreds of well-preserved brachiopods and cephalopods, often pyritized. There is no surface evidence of the fossils inside. Marine fossils found in these concretions are generally not crushed like their neighbors in the shale enveloping the concretions.

Not all large concretions contain fossils. Those uncovered during mining in southern Iowa are full of calcite crystals but no fossils. Eastern coal fields rarely disclose concretions of interest to a fossil collector. But all concretions must form around a nucleus, usually a fossil, so all are worth breaking open. Any reddish or grayish rounded rock found on the strip-mine waste hills is likely to be a concretion.

Small yellow or gray limestone slabs may show up sporadically in strip-mine dumps. These are usually highly productive of marine fossils. Limestone boulders brought down by the glaciers may also appear in the topmost overburden of northern strip mines. These are usually weathered to a yellowish color. Traces of fossils may appear on the outside, but the boulders must be broken apart to release undamaged fossils inside. There

This benevolent little face is not part of a trilobite, but is merely a concretion that happens to resemble a trilobite. Many such pseudofossils are thought to be heads, feet, wings, and bones by overimaginative collectors.

may be some difficulty in identifying fossils and even in determining their age in the stray limestone lumps.

Sandstone layers several feet thick may occur as a blanket deposit in some coal mines. These evenly bedded sandstones may show plant remains, especially where a sudden flood washed in sand that buried standing plants. Such fossils usually lack the fine detail preserved when the plants were entombed in fine clay and mud. However, thick stems, great slabs of bark, and compressed trees can be found in these sandstones as casts.

Masses of crushed *Lepidodendron* and *Sigillaria* are common in a two-foot-thick sandstone layer exposed south of Pella, Iowa. Quantities of leaves are buried at odd angles throughout a thick sandstone near Ottawa, Kansas. Giant stumps and logs of Coal Age trees stand upright in the thick Millstone Grit formation of Indiana and created quite a nuisance during the last century for producers of millstones. A similar occurrence is seen in a roadcut near Omaha, where *Calamites* stems stand upright in the layers of sandstone. Energetic collectors have dug down as much as four feet to remove a long section of stem.

Plant leaves in sandstone are often preserved as a carbon film, and a

Sandstone casts after tree bark are common above a coal seam near Pella, Iowa. The gritty sand does not retain a sharp impression. Inner bark of *Sigillaria*; Pennsylvanian.

spray of lacquer or liquid plastic may be necessary to fix them to the matrix for safe transport home.

Shaft coal mines offer only one place to collect—the dump, which is a mixture of rock from the mine's many levels, most of it from digging the shaft and widening the underground workings. Collectors are not allowed underground, and most active mines will not even allow them on the dump. On an old dump, go around and up and down, breaking concretions and blocks of rock, as well as keeping alert for loose fossils.

GRAVEL PITS

In northern parts of the United States, gravel and sand pits operate in concentrations of glacial sands and pebbles such as kames, eskers, or well-sorted moraines. Elsewhere, and in some parts of the north as well, gravel pits usually lie in old riverbeds or near rivers. Many gravel pits operate in the Mississippi River. Gravel and sand pits are marked on topographic maps, or they can be found in telephone books under the listing "Sand and Gravel."

Usually, the gravel in the piles is sorted as to size, and it is clean, so that fossils are easy to see. The fossils in one gravel pile may have originated from anywhere within a thousand-square-mile area and may be of half-a-dozen geologic periods, with many of the key features for identification worn away. However, a few pretty corals, usually silicified, can be found along with crinoid stems and pebbles that show cross sections of brachiopods, clams, and crinoids. Gravel pits from the Dakotas down to Texas and west to the coast produce an occasional worn piece of petrified wood.

Gravel pits anywhere in the United States, however, may turn up bones, teeth, and tusks of Ice-Age animals. These fossils are much better preserved than the waterworn pebbles of invertebrates and wood because the giant mammoths and mastodons were indigenous, and their bones settled to the bottom of a river or lake with little movement and suffered little damage or abrasion. A pit may operate for years without dredging up a single bone and then for a period of weeks run into masses of bones. The best place to look for these bones is in the office of the gravel pit; the bones and teeth arouse the curiosity of operators and wind up as doorstops.

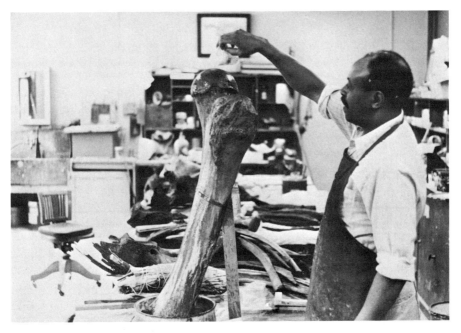

John Harris of the Field Museum pours thinned shellac over a crumbling mastodon leg bone. Large, porous bones such as this will disintegrate when they dry out.

Collecting in a gravel bed west of Fort Pierre, South Dakota. (Photo South Dakota Department of Highways)

The giant gold dredges in Alaska screen enormous quantities of gravel and sand for alluvial gold. Several were forced to abandon one area because of frequent breakdowns when mammoth bones and tusks clogged the machinery.

The bones, which are usually of one of the elephantlike animals, crumble when dried out after being wet for thousands of years. Museums soak the bones in penetrating liquid plastics or shellac, and this should be done by the amateur collector as well.

Tusks found in the continental United States are usually rotted and friable. When removed from the wall of a gravel pit they crumble. Alaskan and Canadian specimens have been preserved by long freezing. The specimens are of sufficiently high quality to have provided a flourishing trade in fossil ivory for many years. The Eskimos used fossil ivory for carving and for awls and fishhooks.

Gravel pits from Iowa westward into the mountain states and south into Texas produce numbers of fossil Pleistocene horse and buffalo teeth. These teeth are several inches long and usually dark brown or black. Mammoth teeth have a platelike structure and may break into slabs that are hard to recognize as parts of a tooth. They are generally white, yellowish, or brownish, and look like ivory. An entire tooth, which can weigh ten pounds or more, is a collector's prize. Mastodon teeth, which

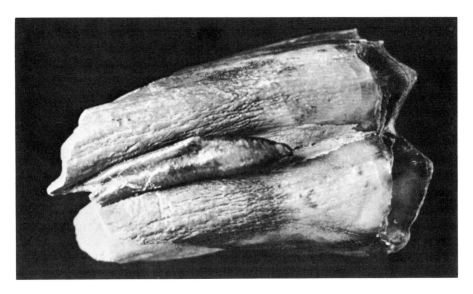

Gravel pits in the West and Southwest are sources of teeth of Pleistocene vertebrates. This bison tooth came from Nebraska.

are even rarer, are readily recognizable as grinding teeth. From time to time bones and teeth of other animals from tiny rodents to giant beavers are found in gravel piles.

METAL MINES

Dumps of iron, fluorite, lead, zinc, and other mines may contain fossils for the collector, because nearly all common metals are mined from fissures and cavities in limestones.

Metal-mining areas are usually well known. The state Geological Survey can often supply a list of operating mines; so can a university geologist. Mines, both operating and abandoned, are marked on topographic maps. Currently or recently operating mines are best for fossil hunting, as old dumps weather rapidly. As with all operating quarries, pits, and mines, the collector should ask permission to hunt, and he should ask where it is safe to hunt.

The state Geological Survey can usually suggest a book or paper about a particular mine that describes ore-bearing horizons, gives their formations and age, and sometimes gives detailed lists of the fossils found in each stratum. This was not done specifically for the fossil hunter but for the convenience of geologists or prospectors wishing to identify the ore-bearing formations.

Hunting is limited to the dumps, where the rocks have been cleaned by rains and fossils have been etched loose by acid waters. Many strata are cut through in working a mine, and fossils on the dump may range from Cambrian to Cretaceous. Midwestern metal mines tend to produce ore from one or two distinct formations, making identification easier.

The best fossil hunting will be found in mines in the flatlands of the Midwest and South; the metamorphism that occurred during mountain building in the East and West destroyed or grossly distorted fossils in the rocks, though there are some undisturbed areas. The iron mines in Minnesota are in Cretaceous rocks that contain excellent fossils. The iron is believed to be a fossil itself, concentrated from the sea water by innumerable bacteria. The seabed that is now the iron mine was the home of clams and snails which can be found in the dumps, turned to dark-brown hematite or limonite.

The lead- and zinc-mining area around Joplin, Missouri, is operating in Mississippian-age rocks that contain characteristic invertebrate fossils. Few mines are still operating, but old dumps are everywhere and contain

Many lead and zinc mines are located in fossiliferous limestone. Their dumps can be productive of specimens of fossils as well as of minerals. Lead mine near Galena, Illinois.

some fossils as well as mineral specimens. Recently, trilobites, cephalopods, and rare new fossils were found in the early Paleozoic rocks exposed at the bottom of barite mines south of St. Louis, Missouri.

The lead and zinc mines, a few of which are still operating in southwest Wisconsin, once covered the area where Illinois, Iowa, and Wisconsin meet. The ore zone here is in Ordovician rock, and dumps have produced fine brachiopods and some trilobites. In southern Illinois and northern Kentucky, the fluorite mines are recovering more lead and zinc than fluorite in Mississippian rocks. Not much waste material is dumped from these mines, but some fossils can be found.

In mines in Pennsylvania producing quartz sand for glassmaking, giant crinoids have been found in the sandstone blocks. They are fragile and difficult to collect but remarkably complete and well preserved for sandstone fossils.

Phosphate strip mines in central Florida, are bonanzas of fossils, because the phosphate is a product of animal life of the ancient Florida seas. The deposits are mostly Miocene, and giant shark teeth, bones of fish and manatee, turtle and alligator plates, and assorted mammal teeth are regularly found in the waste piles. Mammoth teeth and bones of more recent age are occasionally uncovered in the upper levels before the older phosphate-bearing horizon is reached.

ROAD CUTS

Besides mines and quarries, other man-made excavations are probably the best hunting areas for fossils. Fresh material is regularly exposed, and the areas opened for inspection offer much more than is naturally exposed by the slow weathering of shales and limestones. New programs of road building, particularly the interstate highways, are creating road cuts where previously there were none. Roads are widened periodically, and fresh cuts are made. Even old road cuts produce a steady crop of fossils through weathering. Road cuts, except for interstates, are easy places to collect; they are "right beside the road." Interstates are posted with signs warning that no stopping is allowed except for emergency repairs. They may often be reached, however, by driving onto frontage roads and walking down to the road cuts. Some interstates are heavily fenced, with barbed wire on top. Nor are road crews too happy with rockhounds who uproot the grass planted to stop erosion on bare exposures.

Just driving around, particularly on newer roads, is the best way to find road cuts. If an area is known to be fossiliferous, such as around Cincinnati, all road cuts are likely to be fossiliferous wherever shale or

A collector reaches a fossil the previous collector could not reach. Here he uses a twelve-foot pole to dislodge a concretion from a vertical sandstone road cut in Washington. He hopes it will contain a fossil crab.

limestone layers are exposed. Even roadside ditches in fossiliferous areas may be rewarding. Weathering of the limestones releases the fossils and produces topsoil to cover them. Water in the ditches washes the fossils clean and leaves them scattered in the bottom.

Field-trip guides published by state Geological Surveys and by professional geological societies for field-trip meetings often pinpoint prominent roadcuts along the route and describe in detail the formations and fossils exposed. A few states have published roadside geology tour guides for some interesting stretches of highway. One geological society has erected signs along the highway through the Arbuckle Mountains in Oklahoma describing the age and formation of the rocks. State highway departments can usually furnish information about a new interstate highway under construction. Up-to-date road maps often have the interstates under construction designated with dotted lines.

During excavation in 1965 and 1966 for Interstate 71 near Cleveland, Ohio, a bed of Devonian shales carpeted with fossil fish was uncovered. Construction stopped on that stretch while museum crews removed 10,000 specimens, including thousands of fish-bearing slabs. Many were species new to science, and this one chance find added more Devonian

Road cuts often yield fossils abundantly. This one near Bloomington, Indiana, is carpeted with crinoid stems and other Mississippian fossils. (Photo by Betty Crawford)

fish specimens to scientific collections than had ever been collected previously in the United States.

There is little choice where to collect in a road cut. Alongside the road there may be a pile of rubble from weathering of higher layers. It will give a quick idea of what might be found in the entire cut, although fragile fossils will never appear in the weathered material at the bottom.

Limestones in a weathered road cut stand out as prominent ledges. So do some sandstones. Shales weather away, and a road cut through shale becomes a grassy hillside in a very few years. If the shales are sandwiched between limestone or sandstone layers they will weather back, leaving the more resistent layers of hard rock protruding. It is worthwhile digging back into shales to the unweathered part to see whether soft fossils are locked in them.

A small exposure near Vinita, Oklahoma, along the interstate was a popular collecting place for several years. It is now solidly grassed over. The shales on the hillside weathered rapidly, releasing a multitude of Archimedes screws and brachiopods. Few collected the thin layers of limestone that littered the hillside, although these contained well-preserved brachiopods and rare trilobite parts. One Oklahoma collector painstak-

Thin limestone layers of this road cut stand out prominently, while the softer shale crumbles. Both limestone and shale may have fossils.

ingly sampled the hill from top to bottom, digging back into each separate shale layer. He was rewarded by finding a layer near the base which produced crinoid crowns, beautifully preserved. He followed this layer along and found a nest of crinoids that held more than fifty perfect specimens.

RAILROAD CUTS

In the 1800s, railroad cuts were the source of many fossil collections. Roads went up and over the hills rather than through them, and few quarries were operating, so that railroad cuts created the only fresh exposures. A few cuts, mostly those through shales, are still productive. Many can be reached only after quite a bit of walking and are nearly vertical, making collecting difficult. Old cuts through famous fossil-bearing areas are still worth exploring, if for no other reason than that few people still collect from them. Many miles of railroad cuts around Cincinnati produce quantities of Ordovician fossils. So do nearby quarries and road cuts.

Railroad cuts are good places to collect, although most are old and overgrown. This one at Madison, Indiana, has been a famous fossil site for over a century. (Photo by Betty Crawford)

From time to time a beautiful fossil is found in the crushed rock road bed of a railroad. If the material is fresh, it can be traced back to the quarry that produced it. Some railroads still operate quarries to provide their fill. Some are familiar to fossil collectors, such as the railroad quarry at LeGrand, Iowa, which produced the magnificent slabs of Mississippian crinoids and starfish mentioned in Chapter I.

NATURAL EXPOSURES

Fossils will occasionally be found along riverbanks, on mountain cliffs, and in other places where rock is exposed naturally.

Rivers and Creeks

A river or creek draining an area that has exposures of fossiliferous rocks is bound to have a few durable fossils mixed with the gravel in its bed. While rivers and streams do not produce quantities of good fossils, they may provide clues to productive areas in the rocks.

Any detailed map, particularly any topographic map, clearly shows all rivers. Rivers in any part of the United States known to have sedimentary rocks of Paleozoic or younger age are potential fossil sites. (See state maps in Appendix.)

The best time to collect is during low water, usually in late summer. A pair of tennis shoes and a swimsuit make wading a pleasure, though canoes may be necessary on larger rivers to go from gravel bar to gravel bar. Some fossils will be found loose in gravel bars and beaches along the river edge. These will be worn to some degree and may have been transported a long distance by the river or by a glacier, making identification difficult.

The best collecting will be from shale and limestone exposures cut through by the river. Water softens the rock and it also expands while freezing, which helps loosen fossils. Early spring thaws wash down quantities of rock from the riverside exposures. Some riverbank collecting areas of a century ago still produce fossils, although the bank may have moved back hundreds of yards during that time. Flood waters remove loose material from the banks and cuts and expose fresh material for weathering. Collecting is a matter of walking along the river and sampling the exposures in the banks and the river bottom.

Rivers near Cedar Rapids, Iowa, cut through coral-bearing horizons of the Devonian, and some of them are carpeted with worn pieces of coral weighing up to fifty pounds; Kentucky rivers have done the same thing with Devonian and Ordovician corals. The Falls of the Ohio is a site famous for more than 150 years for the quantity, variety, size, and preservation of the corals exposed in the river during low water.

New York rivers in the area west of Buffalo cut through the seemingly bottomless Devonian shales and are all good sites to collect brachiopods, corals, and occasional trilobites. Texas creeks, when they have water in them, flow through Cretaceous rocks, particularly in an area from Dallas south for many miles. The flash floods in these stream beds tear loose and later deposit large echinoids, ammonites, and clams. Maryland and Virginia rivers cut into Miocene and Eocene fossil beds, releasing their modern-appearing shells and redepositing them along the edge of the river.

The first Pennsylvanian plant concretions in the Mazon Creek area of northern Illinois were collected in the bed of the creek. The river is rarely hunted now, because nearby strip mines have uncovered the same concretions, but modern-day hunters still find fossils where collectors hunted in the 1860s. Much of the hunting is done by groping in the muddy water for the rounded concretions.

Western rivers flow through thick beds of basalt that contain petrified

A small creek not only exposes shales and limestones along its banks, but gently erodes fossils from matrix. Near Versailles State Park, Indiana. (Photo by Betty Crawford)

wood, particularly in Oregon, California, Washington, and Idaho. In the John Day basin of Oregon, deep-cutting rivers expose Tertiary beds with rare fossils of nuts and fruits. South Dakota streams in the Badlands wash loose mammal teeth and bones, and in other areas uncover giant concretions containing well-preserved ammonites.

The list could go on and on. Only a few states in the Northeast cannot be expected to produce fossils from at least some riverbeds. Perhaps the greatest exposure of fossil beds in the world is the Grand Canyon, where sedimentary fossil-bearing beds of many geologic periods are exposed by the Colorado River. No collecting is allowed, and even if it were, the 5,000-foot-high banks would be difficult to scale.

Dredging of some Florida rivers produced great quantities of bones and teeth, mostly of Pleistocene animals. Scuba divers have now explored some of these rivers and found underwater caves and small areas loaded with accumulations of bones, which must have been there since the animals died in the cave or in the river many thousands of years ago. Use of scuba gear is a new technique in fossil collecting.

Ocean and Lake Beaches

Hundreds of thousands of miles of coastline border the salt- and fresh-water bodies of the United States. Few beaches are strictly sand, even the famous Florida and California beaches, and some are quite rocky. Fossils can be found in such gravel and rock accumulations, if they have been derived, at least in part, from a sedimentary rock. The fossils are usually poorly preserved, but the collecting is pleasant.

As with river collecting, beach fossils are either waterworn pebbles or fresh specimens from cliff exposures cut by the water. Storms erode beaches, carrying away sand and leaving behind rocks containing a rich harvest of fossils. Winter freezes many northern lakes, and as the ice expands it pushes many feet up the shore, carrying with it rocks once firmly bedded in the bottom.

The coastline of America exposes few fossiliferous cliffs, whereas in England, fossils can be found most of the way around the island. A few isolated coastal cliffs from California into Washington have large exposures of Cretaceous and Tertiary marine fossils. Some California beaches have a few fossils, such as agatized clams and pieces of wood. East coast beaches from Maryland south may yield a stray bone, tooth, or shell. The best-known American location is along the Maryland coast, where large cliffs, rapidly eroding, have released thousands of superbly preserved Miocene shells and some shark teeth.

Little is found around the Gulf beaches except where dredging, particularly from Mississippi almost to the tip of Florida, has dipped into fossil beds lying near the surface .These are primarily Eocene and Miocene with typical shell, bone, and teeth fossils. So many fossil shark teeth are found on the beach at Venice, Florida, and for thirty miles on either side of Venice, that postcards proclaim Venice the shark-tooth capital of the world. During the tourist season it is common to see half a hundred fossil hunters, bent low to the beach, scurrying back and forth with the waves, waiting for shark teeth to appear. During rough weather, large pieces of bone are washed up—freed along with the shark teeth from a fossil bed in the floor of the Gulf.

A small area at the south edge of Tampa, Florida, used to be covered with agatized shells and pieces of coral of Miocene age until the fossil hunters and gem cutters discovered the beauty of these specimens. After the fossil bed was located only a few feet below the surface of the beach, hunters dug up most of the beach out to the low-tide line. Now few pieces remain to be washed up by the waves.

Lake Michigan beaches provide quartz-replaced corals of Devonian or Silurian age at almost any location on the south end of the lake. The

A good collecting spot may be very small. This light-colored hill, barely 50 feet across, is the only exposure near Kebo, Oklahoma, of a Devonian limestone full of trilobite fossils. It is called White Mound.

Petoskey stone found near the Michigan town of that name along Lake Michigan and nearby lake shores is the calcite-replaced Devonian coral *Hexagonaria*. Similar corals are found on the other side of the lake in Door County, Wisconsin.

Lake Erie laps against many Devonian exposures along its eastern half, washing out brachiopods and corals. Canadian shores around Collingswood, Ontario, are paved with Ordovician slabs crowded with trilobites. In one Ohio lake, boulders were found that turned out to be the stumps of petrified Devonian trees.

Many reservoirs and artificial lakes in the West and Southwest have provided access to fossil layers exposed in once-inaccessible river cliffs. There lapping waves concentrate fossils along the shoreline. Lake Texoma, on the Texas-Oklahoma frontier, is a favored spot to wade in the shallow water and feel out large ammonites with the toes. The spillway of Lake Benbrook in Texas is a bonanza of clams and echinoids after water is released, tearing loose many fossils. The spillway of one eastern Oklahoma artificial lake was cut into a layer of limestone composed primarily of crinoids and the blastoid *Pentremites*. Stock tanks in the Southwest are often dug into fossiliferous stone, where the waters clean and expose the fossils.

Calvert Cliffs, Maryland, a Miocene formation where shark teeth are collected. (Photo Maryland Geological Survey)

Other Natural Exposures

There are natural exposures of rock in mountain cliffs and hillsides, especially in the West, that have never been explored by a fossil hunter. Some, such as sedimentary rock caps in the Rocky Mountains at 10,000 and 12,000 feet are nearly inaccessible. Topographic maps give clues to such potential fossil sites: look for steep hillsides and cliffs, which are evident on the maps.

Shark teeth found at Calvert Cliffs. Similar fossils are found in Florida on beaches and in phosphate pits. (Photo Maryland Geological Survey)

Couple viewing fossils protected from elements by plastic shield, at a station along a fossil trail in Badlands National Monument, South Dakota. (Photo South Dakota Department of Highways)

Any location where loose rock is exposed is a potential fossil site, so long as the rock is sedimentary and of proper age. Fossils may erode from the rock and be carried by rains into depressions and ditches. The surface rock on an exposure which is not too steep will probably be black from extreme weathering, and all fossils will have been obliterated. One of these slabs, when turned over, will show a much cleaner side. Turning slabs also exposes scorpions and snakes and should be done with a long-handled pick.

It is often difficult to locate the source of a fossil found near the bottom of a mountain slope or hillside. Soft layers slump over other layers hiding them completely. Pieces roll down the hill and are moved by animals. It is relatively safe to assume that the original location of a fossil was on the slope above the spot where it was found, but perhaps not directly above. A fossil hunter narrows his search as the gold prospector does when he finds a trace of color in a river bed; the prospector follows it upstream until he reaches a point where there is no more gold. The source must lie below that point. Similarly the fossil hunter traces a fossil up the slope.

C. D. Walcott uncovered a slab of shale containing an extraordinary, soft-bodied Cambrian worm while collecting in the mountains of British

Shale formed of volcanic ash entombed these leaves. Oligocene; Florissant, Colorado.

Columbia; he searched for several years before he found the source. This pocket ultimately produced hundreds of slabs of intricately preserved, soft-bodied animals, most of which were new to science.

The folding and severe faulting of western and eastern mountains makes prediction of the location of any rock layer nearly impossible. But it also makes any area potentially worthwhile, even though the predominant rocks look unpromising.

Not far from Walcott's quarry in the Canadian Rockies is a location high up on Mount Stephen where about fifty acres produce an abundance of Cambrian trilobites. The spot is surrounded by unfossiliferous rocks so old that fossils would never be expected anywhere in the area. Geologists believe the block of fossil-bearing shale was torn loose from an unknown location and fell on the side of the mountain during some extensive folding and faulting.

The Cambrian trilobites of Utah are found on the sides of small mountains in the central part of the state. Shark Tooth Hill, near Bakersfield, California, is slowly being dug away for its dental fossils. The Badlands of South Dakota are ideal places to collect vertebrate fossils and some marine fossils. So are the chalk hills of Kansas, which have fossils of marine dinosaurs and giant fish. Hills near Kemmerer, Wyoming, formed of chalky white shale deposited in Tertiary lake beds, contain fossil fish. The foothills of the Rockies around Florissant, Colorado, have numerous areas of soft shale, derived from volcanic ash deposited in freshwater lakes. As the dust settled through the water it buried leaves and insects that are now beautifully preserved in the Oligocene shales.

OTHER AREAS

Any exposure, natural or artificial, is worth looking into. Any area known to have produced fossils, even in the most built-up areas, may at some time have a construction project that will again expose fossil-bearing rocks. Some likely places are described here.

House and office-building excavations

During construction of building foundations in downtown Kansas City a number of years ago, a layer of shale was dug out that was loaded with superb crinoid crowns. It is difficult to collect in excavations for giant office buildings, but the material being removed has to be dumped somewhere, usually as fill. Ask the dump location from any driver as he waits for his truck to be loaded.

Subways and tunnels

Not many new subway lines or tunnels under bays, rivers, and congested areas are being built, but when they are, vast quantities of material have to be removed and dumped. This is also true where eastern and western mountains are being pierced by interstate highway tunnels.

Sewer lines, cable trenches, pipelines

For some large-scale sewer projects, trenches twenty feet deep must be made, and sometimes even deeper cuts must be made for oil pipelines that run for hundreds of miles. Underground power lines and telephone lines, while not buried deeply, may still cause removal of weathered fossiliferous material. Even after the lines have been covered, a mound remains where rain may expose a few stray fossils. Ammonites have come from sewer construction in Colorado Springs. Tons of agatized coral and mollusks came from digging of a five-block-long sewer trench in Tampa, Florida. Texas pipelines rip through miles of Cretaceous fossil beds.

Canals and dredging

The beautiful Silurian eurypterids, now in every major museum, were uncovered during construction of a canal at Lockport, New York, in the last century. Canal building and widening in Chicago cut through Silurian limestones containing trilobites, large cephalopods, and unique carbonized worms. Dredging almost anywhere in Florida is likely to unearth Pleistocene shark teeth, bones, and turtle shell plates along with the sand. Collectors stand at the discharge end of the dredge pipes and grab the dark-colored fossils as they fall in the wet sand. Dredging for ship channels throughout the Southeast and particularly in Florida has left little islands of dredged-up material often containing fossils. Any port city periodically dredges its harbors and may add new channels to its facilities. All are potential fossil-collecting grounds.

City dumps

Old quarries make convenient dumps, and as the refuse rises in the quarry, prolific strata magically become available at waist level. Some dumps accept only clean fill, and some of this fill may be from excavations into fossil-bearing stone. A dump at Alden, New York, recently received a dozen truck loads of shale that turned out to be filled with trilobites. The shale had been removed from a basement excavation.

Harbors and ballast dumps

Ships once carried rock as ballast in their holds. When they reached port, they would dump this in the harbor or along the shore if they were going to take back a full load. A ballast dump near a harbor, particularly along the East Coast, may contain foreign fossils. English ships often loaded up with flint, the type that weathers out of the chalk cliffs of Dover. Some of this flint has replaced ammonites and echinoids, making choice collector's items. Unfortunately, dumped ballast does not carry a label of origin, and it may be impossible to track down its fossils.

Dam construction

Large dams, such as those that have been constructed in recent years in the West and Southwest, are unmatched among major building projects for the quantity of rock they expose during construction. River cliffs are blasted back to the solid rock to give firm anchorage for the concrete wall that will dam the stream. This rubble may be dumped out of the way. Quarries are opened nearby to produce aggregate for the concrete of the dam. These may afford fine collecting. Even small dams will usually be cut into rock. Farmers who scoop out small stock tanks and ponds may also scoop out quantities of loose fossils.

Peat bogs

Peat bogs are a rather rare and specialized type of collecting site limited to the northern glaciated states. Remnants of the last retreat of the ice, these bogs were death traps for Pleistocene animals. Mammoths and mastodons ventured in, became mired, and drowned in the black waters. The high acid content of the peat preserved bones and tusks and teeth. Whenever a peat bog is drained or dug out, there is a good chance of finding vertebrate fossils.

One such bog was scraped down to its clay bottom to make way for an expressway north of Chicago, and the rich peat was dumped on a farmer's field. The first rain washed out an accumulation of bones and teeth believed to be the remains of several complete mammoths. The farmer still plows up bones every spring and has a barn full of fine specimens.

Hog pens and animal burrows

Hillsides may have untold quantities of loose fossils in their soil, weathered from fossiliferous shales and limestones near the surface, but covered by grass and weeds. There is no more helpful animal than the hog for rooting out loose fossils in such places. A hog pen, though not the pleasantest place in which to collect, can contain lots of fossils. One Chicago area club has frequent field trips to a hog pen in Indiana that is paved with large crinoid stems and crinoid slabs showing sporadic plates and even crowns. Such burrowing animals as the badger may go below the plant covering and bring up a few fossils.

Tar pits

The major collecting site of this sort is the well-known La Brea tar pit in Los Angeles. But some day an amateur may stumble onto another tar pit. As in the peat bog, animals became mired in the tar as they came to drink. Predators attacked the helpless Pleistocene animals and themselves became stuck. All sank slowly into the dense oil, which has kept their bones perfectly preserved. Thousands of bones and complete skeletons, even leaves and flowers, have been taken from this tar pit. Sorry, no private collecting.

Caves

Dry areas of the Southwest have a climate that preserves anything that crawls into a cave and dies. A few ice-age animals did just that, such as the extinct ground sloths found lying on or near the surface. The preservation is really mummification, with hair still covering the bones. These extremely fragile fossils are rare enough and ugly enough so that any collector finding one is likely to turn it over to a museum. Caves were also the haunt of ice-age man. No cave should be indiscriminately dug up, lest its scientific evidence be lost.

Fissures

In cave areas where limestone is being quarried, occasional fissures may be encountered. Here joints in the rock have allowed water to sink rapidly from the surface. Such fissures may also have trapped animals that fell into them. A wealth of bones, mostly of tiny Pleistocene rodents, can come from such cracks. Florida has been particularly rich in these fossil-bearing crevices.

Deposits much older than Pleistocene are occasionally found as crevice fillings. The only evidence that northern Illinois was covered by seas more recent than Silurian comes from Devonian fossils, primarily fish teeth, found in a gray shale that filled a deep crevice in the lighter-colored Silurian dolomite. This was exposed briefly during quarrying, long enough for a sharp-eyed collector to sample the unusual shale and discover fossils of an unknown sea. Permian bones, a rich concentration of small amphibians and reptiles, were found near Fort Sill, Oklahoma, in fissures in much older Ordovician limestone.

Any cracks running into what is otherwise solid stone may be filled with fossil-bearing material of a recent age, particularly if the fill is quite different in appearance from any other material in the quarry.

Unusual fossil sites

There are other strange places to search for fossils but hardly common enough to deserve more than a passing mention. In dinosaur country from Colorado and Utah into Canada, occasional piles of gastroliths, or gizzard stones, are found. Some ancient lizards swallowed pebbles to help grind their food, and by definition these stones are fossils. On occasion the gastroliths were themselves fossil pebbles. Nicely rounded and tumbled, these fossils are all of animals older than the Cretaceous or Jurassic.

Ancient man accumulated objects just as the modern fossil collector does. A strange fossil would be picked up and perhaps even buried with the owner. Crinoid stem segments with natural holes in the center were ideally suited to be made into necklaces to adorn the necks of prehistoric men and women. In ancient living sites and graves, fossils collected thousands of years ago can occasionally be collected again.

Fossils are even found in other fossils. The trilobite *Vogdesia* crawled into abandoned shells of large straight cephalopods that littered the bottom of Ordovician seas. Both shell and trilobite sometimes wound up as fossils. Collectors in northeastern Iowa have found that by breaking open large cephalopod shells they may discover an occasional trilobite. Fossil nematodes were found embedded in fossil scorpion skin from Devonian rocks of the northern Rockies. Petrified wood from some areas, when broken apart, reveals the clams known as teredos, still preserved at the end of their burrows. Delicately preserved flowers and leaves from the late Pleistocene are known to science mainly through their unique preservation in the mouths and stomachs of Siberian mastodons and mammoths that were frozen with their mouths still full of food. The diet of Ordovician cephalopods that lived in Arkansas seas has been studied through their stomach contents, now pyritized like the cephalopods themselves. Scales, fins, and teeth of small Pennsylvanian fish are known only from the fossilized excrement of other fish and sharks, never having been found as complete fossils in the rocks.

A remarkable find was made in the petrified stump of a Pennsylvanian tree when it was broken apart. The tree had evidently been hollow, and this hollow was the home of a rare early reptile. The tree and the reptile were fossilized together, the stump protecting the fragile bones from separating and disintegrating. One more proof of what every collector eventually discovers—fossils are where you find them!

VII

PRACTICAL
FIELD TRIPPING

The fossil collector can learn the locations of currently productive fossil sites from other collectors, and from museums, university and state geologists, recent guidebooks, and articles on field trips in hobby magazines. He may also search through professional journals, such as the *Journal of Paleontology*, for those articles that list sites for collecting the fossils described in the articles.

Thus the field trip begins at home, where decisions about destination and itinerary must be made, and skill in recognizing particular fossils must be developed.

PRELIMINARY PREPARATIONS

Several considerations enter into decisions about the destination of the trip. The time of year is important; so is the suitability of the climate for outdoor collecting. Equally important is the collector's particular interest or specialty. He may prefer fossils of one period, such as the Devonian, or of one type, such as trilobites. Desert regions are fit for man or beast only from September through May; high-elevation fossil sites of the northern Rockies are snowbound for about the same period.

Sensible travelers not only make careful plans for their trips, but they also schedule alternative collecting areas. Their first choices may prove to be inaccessible or they may have been used as building sites or dumps; other locations may have been over-collected.

Besides knowing where they are going in such detail as the obscurity of the area requires, good collectors allow ample time to reconnoiter the location. It takes time to discover fresh deposits at a heavily collected location or even to trace out the geology of one that is less well known. Field work is usually preliminary to the actual collecting.

Maps and Guidebooks

Maps, of course, are the traveler's first resource. Information about available maps and directions for understanding and using them appear in Chapter VIII.

After getting the map situation firmly under control, the experienced collector will draw on all available resources to familiarize himself with fossils he expects to find on his trip and the formations in which he is most likely to find them. Many states publish monographs, guidebooks, and popular reports about their fossils. Usually these are listed in pamphlets obtainable from the state Geological Survey or Conservation Department. Even old pamphlets may suggest areas worth a visit although the specific sites are no longer active mines or other workings. State reports of any kind are invaluable, but they become even more useful if their information is enriched with that gained from visits to museums and dealers, where the collector can see and perhaps handle fossils like those he hopes to find. If possible, he should try to see some of the fossils in matrix; this will not only acquaint him with the organism itself but with the rock in which it is found.

Searching out in advance fellow collectors living near the chosen collecting site can bring rewards in information, in specimens, and possibly in new friendships. You can phone or write members of mineral or archeological clubs, museum curators, and college or high school instructors. Often one of these fellow collectors will be glad to help you, and his guidance in the field can be invaluable.

Location markers, road signs, and roads change in our restless society, so that the companionship or advice of a local collector may be needed to bring up-to-date even the newest maps. This may be the margin between a successful trip and disappointment.

Equipping an Automobile

Fossils, like minerals, seem to hide away in the far places of the land where they must be sought after by automobile. Having the vehicle in first-class running order is the next essential for a successful trip. A jeep or other four-wheel-drive vehicle may be necessary to reach trackless spots, especially in the Far West. Not only should the car be fit for the road; it should also be equipped with a spare tire or two, a hand pump, tools for changing tires and for making simple repairs, a roll of chicken wire for traction in desert sands, a tow rope, a spare fan belt, matches, candle stubs, a flashlight, and maps. Since wheel lug bolts are tightened with air wrenches at service stations, it is well for the driver to see whether he could loosen them if he should have to change a tire on the trip.

Some careful and methodical campers advise packing the car, then unpacking and repacking it. In this way they make sure that nothing has been forgotten; that each item is readily available when needed, especially the spare tire, tire wrench, and jack; and that the car is holding up well under its burden. If extremely heavy loads of fossils are expected, overload or helper springs and heavy-duty shock absorbers can be added to the car at some expense, but too much weight can overload two-ply and four-ply tires.

Many collectors are now relying on small motorcycles designed for heavy-duty, off-the-road use to enter wilderness areas. These can be carried on the back of a car or camper. A new race of extremely tough sportsmen's vehicles has recently been born; most of these have tremendous traction through mud, sand, and rocks and are even amphibious, though not well suited for the open road.

Driver failure is as much to be dreaded as car failure. The driver should be sure he is in good health before he changes from the soft physical environment of an office desk to the taxing demands of long drives followed by hours of walking and climbing.

Cans with pouring spouts are needed to carry spare oil, gasoline, and radiator water on long trips off the main road. Put with them cans or plastic containers to hold a gallon of water per person for each day to be spent in desert or dry camp. Cans of fruit juice and a three-day supply of food that is palatable without heating belong in the adventurer's larder.

Emergency Water

In an emergency, it is possible to make the desert yield enough water to sustain life for a while. All that is needed is a six-foot-square sheet of polyethylene or Mylar plastic, the stuff of which painters' drop cloths are made. A hole about three feet across and twenty to twenty-four inches deep is dug in the sand. A pail or quart can is placed in the bottom of the hole. Over it the plastic sheet is draped, and a small stone is placed with its lowest point over the collecting cup. Stones around its edge anchor the sheet firmly. The heat of the desert sun will vaporize some moisture from the sand—it will collect on the underside of the plastic and drip into the cup. Crushed cactus or other vegetation can be scattered in the bottom of the hole to provide more moisture for evaporation.

Clothing

The weight of clothing proper to the trip will depend on the conditions likely to be encountered. But, rain or sun, cold or hot, the body should be covered: clothing is a necessary protection against sunburn, abrasions, and flying splinters of rock broken off by the collector's hammer. Sharp flakes such as these can blind an eye or cut an artery of the face or arm. Anyone who has seen a fellow collector streaming blood from this kind of cut has had sufficient warning to protect his body in the future. And anyone who has experienced the agonies of sunburn knows it is disabling.

Cotton or wool slacks for men and women; army-last high shoes with molded rubber soles, or good work boots well broken in and worn over two pairs of socks, one thin and one heavy; sturdy work gloves; and a hat or cap with a brim make up a sensible field costume. Clothing should be loose—large enough to allow easy movement in climbing and digging. Sunglasses and suntan oil will temper the brightness of midday, and safety goggles, safety shoes, and hard hats should be available for those occasions when work will be done among heavy rocks or along quarry walls. Federal regulations now require them for anyone collecting in a quarry.

Although they are recommended in many books, high boots are too heavy for most field use. Exposure to snakes would seem about the only excuse for wearing them. Rubber boots may occasionally be needed to collect in streams, on beaches, or in deep mud. The most effective way to adjust to the changing temperatures characteristic of mountain or desert regions is to have several weights of sweaters on hand. A lightweight nylon ski jacket gives excellent cold-weather protection. A pocketknife, a pocket first-aid kit, and insect repellent will be helpful on unexpected occasions. Soap for washing up after handling rocks should be included. Some minerals leave poisonous residues on the hands.

SAFETY PRECAUTIONS IN THE FIELD

The perils and hazards of the great outdoors have been grossly exaggerated. The collector is safer clinging by hand on a cliff than he is in his car on a crowded highway. He will avoid most hazards if he will do three things:

1. Look over the collecting area with an eye to possible danger spots.
2. Keep a level head and rely on common sense.
3. Leave behind word about where he is going and when he may be expected to return. If he fails to return on time, rescuers can go looking for him.

Going alone into the mountains or desert is not common-sense behavior. A sudden storm or even a crippling injury is not half so terrifying when someone is with you as it is when you are alone.

The sensible collector learns to hasten slowly and gradually. He walks with a rhythmic step, putting each foot down firmly so that a loose stone will not mean a sprained ankle. He finds the easiest path and the most gradual slope. If he gets out of breath, he stops and rests, realizing that he is a sedentary worker suddenly turned outdoorsman, not an Olympic athlete, and that unusual exertion at high altitudes makes severe demands on a body unaccustomed and unacclimated to it.

Climbing and walking on difficult ground are much easier if the collector learns to control his body movements so that he is never off balance. Nasty falls can be avoided if, while climbing on rocks, he tests footholds and handholds before putting his full weight on them.

If there are several people in the party, the pace should be set by the physical ability of the slowest. If two or more collectors are together on stony slopes, they should stay close by one another so that dislodged stones will not endanger anyone lagging behind.

Collecting in Dangerous Areas

Common sense teaches that no specimen is worth endangering life or risking serious injury. It is especially important to remember this when collecting in a quarry, where the rock is shattered and may overhang the face. This is the place to wear a hard hat. If pebbles are falling from the overhang, think twice before risking the chance that larger rocks may fall, too. Particular care must be taken in the spring, when frost may be holding the rocks together. A warm sun will soon melt the ice and may release tons of rock. Even a hard hat will not help then.

Open mine shafts, passages to mines, and deep prospect holes can be deadly. They may harbor dangerous animals; snakes retreat from the heat of the sun into the coolness of such places. Falling rocks and rotting ladders make shafts unsafe, and old drifts present the perils of falling roof rocks, hidden pits, and treacherous floor timbers, besides the ever-present possibility of deadly gases.

Collectors have been drowned in the bone-dry desert by flash floods that fill draws and canyons to the brim within minutes. It is advisable to avoid camping in such places.

Getting Lost

Getting lost is perhaps the greatest hazard the collector faces. The desert is a place of austere beauty and of harsh extremes. The mountains turn a forbidding face to the person unfamiliar with them, but it is possible to live, and live well, in both environments. If you get lost, don't panic. Stay with your companions, take shelter where you are, and try to enjoy the situation as you wait for rescue. Prospectors in such situations have been known to live on fern shoots, pinon nuts, and the berries that bears eat, such as rose hips, elderberries, thimbleberries, and juniper pips.

The lost and lorn who have their automobile have everything needed to sustain life—shelter, heat, water, fuel for making an alarm fire. Desert-wise persons advise staying in and close to the car; it is easier for rescuers to find a car than a person. Given enough reserve food and water, the misplaced prospector can settle down in his steel shelter, safe from animals and weather, and enjoy being away from it all for a few hours or days.

Poisonous Plants and Insects

Woodcraft will enable the collector to identify and shun poison ivy and poison oak. Anyone who is sensitive to their irritants should learn in self-defense to recognize these plants. Washing with soap as soon as possible after exposure is the recommended remedy.

Sunburn and windburn are among the minor perils of the outdoors. Much discomfort can be avoided if the body is covered. Should one be burned, however, baking-soda solution will ease the discomfort, and the moist soda will also relieve the sting of insect bites. For most other purposes, such as cuts and scrapes incidental to all contact with nature, the best medicine is warm water and soap.

Wood ticks and most other insects can be destroyed with sprays. The wood tick would be a minor nuisance except that it sometimes carries the virus of Rocky Mountain fever, a once-deadly disease that now can be controlled with drugs. Ticks are small, flat gray insects that appear in warm weather. They hang on the twigs of underbrush, waiting to hitchhike a ride on a passing animal or man. Once having found a host, they burrow underneath the skin to feed on blood. If they are not too deeply buried they can be removed with tweezers; they can even be made to back out at the touch of a heated nail or a lighted cigaret. If they are beyond the reach of such remedies it is advisable to have a physician remove them, so as to avoid infection of the wound.

Chiggers infest prime collecting areas of Kansas, Oklahoma, and Arkansas and seem particularly fond of eastern city-dwellers who have never experienced their long-itching bites. Insect repellent should always be applied when you are crossing grassy or weedy areas in summer. Dusting clothing with powdered sulfur is recomended to keep both chiggers and ticks away. Lozenges of a sulfur preparation to be eaten for the same purpose are on the market.

Poisonous Animals

The possibility of encountering scorpions—especially the virulently poisonous one of the Southwest—makes it prudent to take certain precautions in regions where these animals are not uncommon. Careful campers do not sleep on the ground, and they shake out footwear before putting it on. Because scorpions, small, straw-colored, spiderlike animals, usually spend their days beneath rock slabs, pieces of rock should be overturned with a pick, not by hand.

Snakes bite a few thousand persons a year, but few of their victims die. Certainly the most dangerous are the various species of rattlesnakes, because they are found in almost every state. Some are quite large and inject a large amount of venom when they strike. The coral snake of Florida is dangerous but is restricted to a small region. Copperheads and water moccasins prefer swamps and damp areas.

Snakes, like all reptiles, cannot adjust their bodies to changing temperatures, so that they are forced to hibernate in cold weather and to hide from the sun in the heat of the day. In the desert they hunt for rodents at night. Few snakes are aggressive; they will bite only if they are driven into a corner where they feel they must fight to escape. For this reason, most encounters with snakes are the fault of man; they can be avoided by taking a few simple precautions.

Avoid walking in dense brush or near fallen logs or woodpiles where snakes may be resting during the day, or anywhere that you cannot see where you are stepping. Be careful of narrow, rocky ledges where you may find a snake sunbathing. Walk slowly and make some noise as you walk to give warning of your approach. Stout shoes and loose, floppy trousers will protect the most exposed area of your body—the legs and feet. Don't reach into a hole or turn over rocks with your hands; use a stick or rock hammer. Stay as close as possible to your car; it may be your ambulance if you are bitten.

Dr. Clifford C. Snyder of the University of Utah, a recognized authority on treatment of snakebites, advises that venturesome persons going into snake country should carry a snakebite kit and learn to use it. Dr. Snyder has permitted us to include his recommendations for a kit, which should include a flat elastic tourniquet, two surgical prep sponges saturated with alcohol and sealed in sterile foil, a disposable scalpel sealed in sterile foil, and an ampoule of antivenin.

The best treatment for snakebite is to rush the victim to a hospital or physician. Meanwhile, the victim should sit down, quiet himself, and avoid exertion. If possible, someone should kill the snake for identification purposes in choosing the correct antivenin for treatment. A tourniquet—even a belt, tie, handkerchief, or strip of cloth—should be applied between the bite and the heart. On an arm or leg, place the tourniquet two or three inches above the bite and above the swelling. It should be loose enough so that a finger can easily be inserted beneath it. If it is too tight, it will stop the circulation. The tourniquet should be loosened once an hour. But if the bite is on the face, do not attempt to apply a tourniquet.

The second step in field first aid is to cleanse the wound with the alcohol sponges. Then an incision should be made, connecting the fang marks and extending the cut a quarter inch beyond them. Care must be taken to avoid severing muscles or nerves. The wound should be squeezed gently with the fingers for twenty to thirty minutes or until the victim is taken to the doctor. Dr. Snyder advises against sucking the wound because the mouth contains bacteria and because unnoticed cuts in the mouth may absorb enough poison to do in the would-be doctor. Ice wrapped in a bit of clean cloth may be applied to the wound, but not for more than an hour. The ice should be removed gradually to avoid stimulating the circulation.

Directions for administering the antivenin should be followed closely if the situation requires that it be given in the field.

Food Poisoning

Food poisoning is perhaps a greater danger than snakes. Prepared meats and sandwich fillings, salad, and cream-filled bakery goods are foods likely to incubate poisons when left in the hot trunk of a car or on the picnic table in the sun. Precautions are doubly important where refrigeration facilities are makeshift. One tainted sandwich is enough to spoil a year's vacation.

THE ETIQUETTE OF COLLECTING

But Mother Nature is fairly kind to those who use her resources wisely and respect her laws. They can enjoy her weather and derive health from contact with her. Similarly, the rights of other men must be respected by the sensible collector, who recognizes that he shares the bounty of nature with others.

Collecting, therefore, has not only its craft but also its etiquette. Essentially, etiquette is respect for the rights of others. It is thoughtfulness, good manners, as well as the selfish fear that collecting places will be closed if collectors abuse their privileges.

And closed they often are. Here is a report from an eastern geologist about the consequences of greed and thoughtlessness in his state:

One of our Coastal Plain localities, which is accessible and for years has been a wonderful source of material for students, has been closed the past three years because a bus-load of students from a relatively new college, with apparently a few new and untrained instructors, walked abreast across a wheat field just before harvest time, in order to get to the ravine where the fossils occur. The more experienced and thoughtful visitors had always taken a more difficult and sloppier road, waded down a stream, got mud in their shoes, but retained the goodwill.

In another instance, the nearest cave for exploration welcomed organized groups for many years. About five years ago, a Philadelphia group broke into the house to get the key to the door into the cave. When the owner returned, she ordered the immediate sealing of the entrance with cinder blocks. It is almost beyond understanding that intelligent and conservation-minded people can become so thoughtless and self-centered. However, this type of activity seems to be on the increase; perhaps because so many more of us have leisure which can be used in the great outdoors.

Several of the better localities, where specific directions have appeared in print, have been completely obliterated. One of these is the so-called Trilobite Ridge in northern New Jersey. There is hardly a piece of rock in the area as big as a quarter. In another location in Camden, a fifteen foot cutbank has been completely levelled, not by the owner, but by fossil collectors. The owner now has a nice, gently sloping lawn.

Incredible as these cases may seem, the prize has been taken by some mineral collectors who dug carnelians under the foundation of a high-tension tower. The area is still available to the knowledgeable collectors only because one mineral club, finding the excavation, filled it and advised the owner.

Elsewhere in the same state, a guest of a mineral club cut his leg while collecting, went to a hospital, and returned to collect personal damages from the owner of the site. Since then, collecting has not been allowed in this locality.

A Code of Ethics

An Emily Post for paleontologists might include the following:

1. Get permission to enter and to collect on private or public property. Most public property is open, but it may be closed because it is under a mining claim or has been withdrawn from unrestricted use for some other purpose. Finding the owner of private property may require some investigation, because property changes hands frequently in this restless age.

2. Do not drive through or otherwise damage growing crops. In farming country it is a good policy to stay on the roads. Do not cut trees or bushes unless permission to do so has been expressly granted.

3. Do not touch machinery, such as that used in a quarry or in highway construction, and be careful not to drive over power cables or air hoses. Be especially careful not to drop any rocks down freshly drilled blasting holes in quarries.

4. Do no blasting without express permission of the owner of the property. For your own safety and that of others, leave the use of explosives to a skilled person.

5. Do not carry or discharge firearms. There can be exceptions to this statement, but they should be legally and logically justifiable. Mixing hunting and collecting trips is likely to lead to trouble.

6. Take every precaution against starting a grass or forest fire. Extinguish and bury the embers of campfires and be careful about discarding lighted cigarets and matches.

7. Do not foul creeks or wells. Residents and livestock may depend on them for water.

8. Fill in any holes you may have dug lest cattle or other domestic animals fall in and injure themselves.

9. Bury, burn, or carry away your litter. Leave a clean campsite.

10. Open gates should be left open; closed gates should be closed behind you.

11. Collect only sufficient material for your own needs. Don't collect to sell. To do so is to turn a hobby into a business and to take advantage of the owners (public or private) who have allowed you to collect.

12. Don't try to carry a load that is beyond your strength, especially on mountain slopes and rocky flats. Control of the body is lost to some degree under such conditions, and you expose yourself to the danger of falls as well as of heart attacks. Neither makes the trip more enjoyable.

13. Do not carry rocks or tools loose in the car. In a collision or sudden stop, they become lethal projectiles.

The Law of Trespass

The fundamental law regulating any outdoor activity such as fossil collecting is the common law of trespass. Trespass was originally a wrong committed with force and violence against a person or property, but in one form of the old writs it became trespass or breach of the close, which meant unlawful entry on another person's property. Thus permission to enter and further permission to camp, excavate, etc., should be obtained. Inasmuch as an owner can be held responsible for injury to a person given permission to enter his land, the owner may require you to sign a waiver to relieve him of this responsibility. This is generally not true if he charges a fee.

Privately owned land is protected against trespass whether or not it is enclosed or cultivated. As a California official said, rural residents tend to be very conservative about property rights, but most of them are reasonable people, and, if the visitor is courteous, considerate, and asks permission to collect, he will be favorably received. "Always obtain permission before entering private property and never leave a mess behind," was his advice.

Owner-Release Laws

Some states have passed what are known as owner-release laws which absolve the owner of the land of responsibility for the safety of persons he has permitted to enter his property. In the form used in Texas, which is representative of many others (except that some restrict the wording to hunters only), the law states that the owner or lessee does not assure the guest that the premises are safe and that he does not take responsibility for the acts of the guests if they injure others. The law does not relieve the owner of responsibility for his own malicious or deliberate acts, nor does it apply if he uses the land for commercial recreation or charges admission. The Texas act limits its application to hunters, fishermen, or campers, but it might be extended to rock collectors, too.

Common decency and good policy alike suggest that some form of appreciation be shown for the owner's kindness toward the collector. This may take the form of a gift, such as a piece of jewelry or a book, or taking the owner out to dinner. Each collector represents all collectors by his public behavior. If he is selfish or inconsiderate, all collectors will be judged to be like him, and all collectors will suffer by finding collecting areas closed to them.

The Antiquities Act

The basic law governing collecting on federally owned lands is the Antiquities Act of 1906 (U.S. Code Sec. 34 Statute L-225). This provides that permits to examine ruins, excavate archeological sights, and to collect objects of antiquity or of historical or scientific interest on lands under their jurisdiction may be granted by the secretaries of the Interior, Agriculture, or the Army to institutions they deem properly qualified, provided that these activities are for the benefit of reputable museums, colleges, or other recognized scientific or educational institutions and that the gatherings will be permanently preserved in public museums. The penalty for violation is a $500 fine or 90 days in prison.

The thought behind the law is that the federal government has the responsibility to protect antiquities that are the property of all the people, and the act has been interpreted to include vertebrate paleontological remains as antiquities. Fossils have rarely been the subject of litigation, but in a South Carolina case, in an application for a mining claim filed to protect collecting on a fossil site, a fossil bone was refused classification as a mineral.

In general, the federal antiquities law has not been used to restrain invertebrate fossil collectors. In 1966, the federal government by a regulation of the Department of the Interior set a daily bag limit of 25 pounds plus one large single piece on noncommercial collectors of fossil wood and an annual limit of 250 pounds a person. The regulation forbids commercial collecting or the use of power tools, such as bulldozers, to unearth the wood. This regulation does not apply to privately owned land, only to federal lands under the jurisdiction of the Department of the Interior.

Rules for Public Lands

More recently, the Bureau of Land Management of the Department of the Interior announced rules under which collecting is permitted on lands under its jurisdiction. The citation, Paragraph 6010.2, Rules of Conduct, of Rules for Public Lands, reads:

(a) Permitted Activities:
(1) Collecting-hobby specimens. Flowers, berries, nuts, seeds, cones, leaves and similar renewable resources and non-renewable resources such as rocks, mineral specimens, common invertebrate fossils and gemstones may be collected in reasonable quantities for personal use, consumption or hobby collecting. Limitations on this privilege are contained in Paragraph (b) of this section.
(b) Prohibited Activities:
(5) Gather or collect renewable or non-renewable resources for the purpose of sale or barter unless specifically permitted or authorized by law.
(7) Use motorized mechanical devices for digging, scraping or trenching for purposes of collecting.

You may wish to send for Circular 2147 of the Bureau of Land Management, Department of the Interior, which reprints laws and regulations affecting mineral and fossil collecting.

THE PUBLIC LAND LAW REVIEW COMMISSION

The direction of federal thinking and probably of future law on the use of federal lands is contained in the report of the Public Land Law Review Commission, which made a five-year study of laws governing the third of the nation's area that is federally owned. Most of these lands are in the western states and Alaska. The Commission recommended that the United States retain these lands instead of trying to dispose of them as in the past, and it also recommended that a policy of classifying lands by their dominant use be adopted. Furthermore, it favored charging for use of public lands; for instance, it recommended a fee for recreational purposes, which would include collecting.

The Commission further recognized "hobby mineral collecting" and recommended that it be permitted on the "unappropriated public domain" under regulations set up by the Secretary of the Interior, who would oversee permit requirements and fees to be charged. Presumably this would permit collecting for personal use on any public lands where it would not interfere with a designated dominant use, such as grazing, mining, farming, or lumbering.

State Laws

Some state regulations and laws also apply to collecting. Below is a summary of those reported in force most recently:

ALABAMA—No specific regulations apply to fossil collecting.

ALASKA—State reserves mineral and fossil rights when selling or granting state-owned lands. "The right to extract fossils (presumably commercially) must be leased from the state."

ARIZONA—Private interests own 16 percent of land, to which trespass law would apply; United States owns 71.1 percent, to which Antiquities Act and petrified wood rules would apply; and for the remaining area, state has its own antiquities act which restricts wholesale collecting.

ARKANSAS—No specific regulations applying to fossil collecting. Landowner relieved of responsibility for injury to person he has admitted to his land.

CALIFORNIA—Much of state public domain, but because of major farming interests laws against trespass are interpreted strictly. State has no specific laws restricting fossil collecting. Digging or carrying away of a stone without permission is specifically defined as an act of trespass.

COLORADO—No applicable law reported.

CONNECTICUT—Owner may register land for recreational use and become exempt from responsibility for injury to person entering land.

DELAWARE—No specific laws except for trespass on private lands, but the U.S. Corps of Engineers objects to wholesale collecting along the Chesapeake and Delaware canal as a theft of federal property.

FLORIDA—The state's regulations for preserving objects of historic or scientific value include fossils, but the laws have been set up primarily to control treasure hunters and have apparently not been enforced against amateur collectors.

GEORGIA—No specific regulations that apply to fossil collecting.

HAWAII—Few fossils exist and no regulations about them have been passed.

IDAHO—Vertebrate fossils may be excavated or removed from the state only by permission of the state historical society under regulations for preservation of historic sites.

ILLINOIS—State retains rights over aboriginal remains and sites but has no specific regulations regarding collection of fossils.

INDIANA—State has law against damaging cave formations but none regarding collecting of fossils. Landowner is exempted from responsibility for injury to anyone entering his property, by permission or without it.

IOWA—Restricts collecting only in parks and other state-owned areas.

KANSAS—Permission of the antiquities commission is needed to dig for archeological objects, but apparently this has not been extended to fossils.

KENTUCKY—The department of archeology at the University of Kentucky issues permits for archeological digs on public lands, but apparently this has not been required of fossil collectors. Owners are released from liability for damages arising from injury to a hunter on their lands, with or without permission.

LOUISIANA—No specific regulations apply to fossil collecting. Owner or occupant of land has no responsibility to keep premises safe for anyone who enters on it, with or without permission.

MAINE—State has few fossils, and no laws applying to fossils.

MARYLAND—Ownership of antiquities belongs to state, and permission to excavate them is given by the Geological Survey. No laws apply specifically to fossils. Landowner is not held responsible for safety of premises or injury to hunters; presumably this would apply to fossil collectors, too.

MASSACHUSETTS—State has landowner's release law for hunters and fishermen.

MICHIGAN—Restricts excavation of antiquities or aboriginal material, but consent of landowner is enough for removal of objects from privately held land. Trespass is defined as willful carrying away of any stone, etc., worth more than $5 without permission of landowner. Landowner is not responsible for safety of persons entering his land for outdoor recreational purposes.

MINNESOTA—No specific regulations apply to fossil collecting.

MISSISSIPPI—State exempts individuals making natural-history collections for scientific purposes from regulations on excavation of objects of historic or scientific value. Permits are granted to qualified persons and institutions.

MISSOURI—Restriction on fossil collecting exists only in state parks, and fossils may be collected there with permission of the park superintendent.

MONTANA—Antiquities law is patterned after the federal act. The state attorney general has ruled that it can apply to fossil collecting but has not been so applied. It might be applied against wholesale collecting for profit.

NEBRASKA—Ranchers have obtained a strict trespass law which does not require them to post signs, but owner of land is relieved of liability for safety of persons on his land for recreational purposes, either with or without permission.

NEVADA—The State Museum may post petrified wood sites against collecting to preserve them. A permit is needed from the state on state lands and from the federal government on federal lands to excavate prehistoric sites, which are defined to include paleontological deposits. But collecting of gems, fossils, or artifacts is allowed without permit if the collecting is not done on a prehistoric site.

NEW HAMPSHIRE—State has few fossils and no laws regulating collecting.

NEW JERSEY—The state has no law restricting the collecting of fossils, but it assumes good behavior from persons permitted to enter on land for such purposes. The state has a law relieving landowners of any responsibility for making certain their property is safe for persons entering on the land for recreational or sports purposes.

NEW MEXICO—Science commission supervises collecting, and a system of state monuments has been created to preserve sites of scientific value. Permits to collect from state monuments must be obtained from commissioner of lands. Apparently no state regulation of other areas.

NEW YORK—Collecting permits on state-owned lands are issued to persons "who are professionally competent" by the assistant commissioner for State Museum and Science Service, State Education Department, Albany, 12224. Permits are good for a year.

NORTH CAROLINA—No specific regulations apply to fossil collecting but do regulate collecting of artifacts.

NORTH DAKOTA—Qualified persons and institutions are permitted to excavate paleontological material for a $5 permit fee, but all material from state-owned land must be given to state. Landowner may collect on his own land or give other persons permission without a state permit.

OHIO—No specific regulations apply to fossil collecting.

OKLAHOMA—Only restriction is to get permission to collect on private land.

OREGON—Persons allowed to make natural history collections for scientific purposes, but permit should be obtained from division of state lands to collect on state lands. Regulations may also be prescribed for collecting petrified wood or gemstones on state lands, and collectors may be required to return as much as one-quarter of what they collect to the state. All beaches and recreation areas are open to public use. Landowners may not be held responsible for safety of persons entering on their lands.

PENNSYLVANIA—No specific regulations apply to fossil collecting.

RHODE ISLAND—Has no landowner's release law and has no state agency concerned with its few fossil deposits.

SOUTH CAROLINA—No specific regulations apply to fossil collecting.

SOUTH DAKOTA—Permit is required from the state historical society to make archeological excavations on public lands.

TENNESSEE—Owner has the responsibility for making his land reasonably safe for invited guests but has no responsibility for trespassers.

TEXAS—Landowner may not be held responsible for safety of persons to whom he gives permission to enter on his property.

UTAH—Casual collecting of petrified wood, fossils, and gemstones in small quantities from unrestricted state or federal lands is permissible if the material is for personal noncommercial use. Otherwise a permit is required. Small quantities are defined as ten pounds or less a day or one hundred pounds a season. No collecting is allowed in national monuments or parks or state parks, or Bureau of Land Management or local recreational areas. Major excavation of paleontological deposits on state or federal lands requires permission of the state park and recreation commission, the county board, and, if on federal land, the federal agency concerned.

VERMONT—No specific regulations on collecting of fossils.

VIRGINIA—No specific regulations on collecting of fossils. A landowner does not assume liability for the safety of persons he admits to his property.

WASHINGTON—Recent law frees a landowner of responsibility for injury to persons using his land for recreation with his permission, provided that he posts warning signs in any potentially dangerous area.

WEST VIRGINIA—No specific regulations on collecting of fossils. A state antiquities commission regulates collecting on state lands, but apparently its jurisdiction does not extend to amateur collecting for personal use.

WISCONSIN—State's trespass law relates primarily to hunters and fishermen. No specific law regulates collecting of fossils.

WYOMING—Collectors of fossils and artifacts from state-owned lands are required to get a permit from the commissioner of lands at the State Capital in Cheyenne. Collecting is permitted only for scientific purposes.

Essentially, then, a few states have regulations on fossil collecting, but even they make relatively little attempt to deny the amateur the right to gather surface fossils for his own use on public land not otherwise dedicated. None of the states attempts to keep jurisdiction over privately owned lands, although a few reserve mineral rights on lands sold to individuals. A growing number of states are making it easier for an owner to give permission by their owner-release laws, which relax barriers erected by insurance companies which might have to satisfy claims arising from an accident.

Roadside collecting may be prohibited by state police, particularly on interstate highways, where roadside parking is usually limited to emergencies. It would be wise to ask the state police whether collecting is allowed in road cuts if the car is parked on another road or in a designated rest area.

VIII

MAPS AND HOW TO
USE THEM

A map, for the purpose of an amateur collector, is a generalized representation on a flat surface of some aspect of the surface of the earth. It may be a road map, such as those that filling stations give away; a topographic map, which expresses a third dimension through such a device as contour lines; or a geologic map or other more specialized type that shows the nature of the surface rocks, or economic and geographical features.

ROAD MAPS

In this automobile age, the collector will, of course, find road maps invaluable. They direct him to where he hopes to collect by the shortest, fastest or most convenient route. A few even indicate paleontological sites such as petrified forests, and fossil fish beds.

But the serious collector will wish to have more precise and analytical information than he can get from a road map or atlas. It is not enough to be guided to a site; the collector needs to know more about the deposit in which he plans to collect. Much of this information he can learn from maps before he ever takes to the road, and there is always more to learn when he gets to his destination. It is rarely possible to drive right to an undepleted fossil-collecting site, and locating the deposit itself usually calls for hiking guided by a good map.

TOPOGRAPHIC MAPS

The collector's best guide is the topographic map. Road maps do not show the steepness of hills, the depth of valleys, or even subtle changes of elevation in the prairie states. The topographic map adds the illusion of this third dimension by conventional symbols known as contour lines. These show the relief of the landscape, which is the difference of elevation of hills, valleys, and other natural features. Thus they show height, which is the difference of elevation of two nearby objects, and elevation, which is the height figured from a base plane, usually sea level, known as the datum plane.

In addition, these maps mark the position of man-made features, such as roads (including minor trails not shown on road maps), mines, quarries, and towns. In the topographic maps produced by the United States Geological Survey, water and ice, such as lakes and glaciers, are shown in blue; man-made objects and political boundaries are in black; major highways are in red; forested areas in green; and the contour lines are in brown. Other symbols are explained on the back of some of the maps, and a summary of the symbols is contained in the pamphlet "Topographic Maps," which may be had free from the Map Information Office of the Geological Survey, Washington, D.C. 20242.

Topographic maps are made to scale, that is, a distance on the earth's surface is portrayed by a proportionate distance on the map. The maps use several scales, but the most common are 1 inch to 62,500 inches, or roughly 1 inch to 1 mile; 1 inch to 24,000 inches; and 1 inch to 250,000 inches. The first is a medium scale useful for rural areas where a great deal of detail is not necessary; the second, a large-scale map for highly developed areas, shows individual buildings; and the last is useful primarily to cover an extensive area in one map.

The scale is shown in graphic form on the margin of the map. By marking off a distance on the map on a piece of paper, the user can compare the distance with the scale and get a reading of the distance in miles or feet. Maps are always made with the north direction at the top. Symbols on the margin indicate the true north and the magnetic north, which are usually several degrees apart, depending on location.

The United States Geological Survey has been mapping the United States and Puerto Rico since 1882. In Canada, similar maps are made by the Geological Survey of Canada, in Ottawa.

Interpreting the Topographic Map

Interpreting the topographic map, like interpreting poetry, takes some practice and some imagination. The map's contour lines are like steps, with a fixed height of "risers." The distance between the contour lines may be 20 feet or 100 feet; the interval is chosen to present a readable representation of the detail involved on the particular map, and the interval is designated on the map.

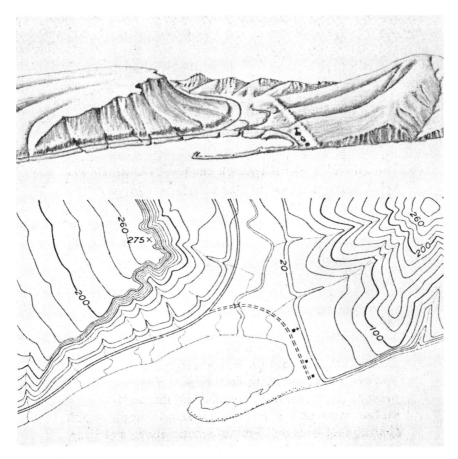

River valley and hills (upper drawing) and topographic map of same area (lower drawing). The river flows into a bay partly enclosed by a sandspit. The hill on right has a gradual slope; the one on left rises steeply above a tableland. From the improved valley road, a dirt road takes off right to a church and two houses. Contours on the map are 20 feet apart. (From U.S. Geological Survey's "Topographic Maps.")

One way to understand the idea of contours is to visualize a view of the land from an airplane. (If one flew over an area on which the actual contour lines were marked on the landscape, he would, in effect, be seeing a contour map, but contour lines seldom occur naturally.) Imagine a series of stakes driven on a hill: first a series at every point with an elevation of 500 feet above sea level, then another row at 510 feet, and so on to the top of the hill. From the air these rows of stakes would appear to form irregular circles some distance apart.

The same thing could be accomplished by creating a giant lake with its surface 500 feet above sea level. Then the shoreline of the lake would be the 500-foot contour line on the aerial map or on a contour map. Raising the lake level 10 feet would establish the 510-foot contour line. In relatively flat country such a 10 foot rise would extend the shoreline hundreds of feet horizontally; the contour lines would be far apart on the map. Conversely, a lake wedged between almost vertical cliffs would show very little increase in size as it rose. Its contour lines would be almost on top of each other.

Each contour line, therefore, represents a series of points of like elevation that form an irregular circle. A hill is a stack of closed loops; a valley is a V-shaped or U-shaped series of lines that cross the valley; and a cliff face is a punched-together mass of lines. If the cliff is exactly vertical, the lines coalesce into one line. Contour lines may run off the map, but if the map were big enough it would be seen that they eventually form a closed loop.

Every fourth or fifth contour line, depending on the scale, is printed darker and carries a figure of the elevation in feet. The elevation of certain prominent objects such as mountain peaks, bench marks, or lakes is given in black. A bench mark (BM on the map) is a real marker placed on the earth's surface that records one of the thousands of points whose elevation has been precisely determined by survey. Usually it is a metal plate on a concrete post. These markers should never be marred or disturbed.

Depressions are also marked with contour lines. Short ticks (hachures) at right angles to the contour lines point down the slope. They are added when there is possibility of confusion between a small hill and a shallow depression.

Topographic maps are periodically revised, especially those where urban growth has caused major changes in manmade structures. Sometimes, however, the collector will wish to find an old fossil location by an obsolete description, such as "one-half mile south of Jones's Ferry on the west bank." Jones's Ferry is long gone. An old topographic map, however, may be found in a library or museum, and it may carry the old site description. By such research, old collecting sites can be rediscovered. The date of a topographic map appears in the lower right-hand corner.

Geological Survey Quadrangle Maps

The quadrangle maps, known as the national series of the U.S. Geological Survey (and those of the Canadian Survey), are based on the meridians of longitude and the parallels of latitude. The meridians of longitude run north and south, dividing the earth into wedge-shaped pieces east and west of the prime meridian at Greenwich, England; the parallels of latitude run east and west and divide the surface north and south of the equator. The United States is in the north latitudes and west longitudes.

The maps are printed on large sheets of paper. Each of the large-scale 1-to-24,000 maps covers $7\frac{1}{2}$ minutes of latitude and longitude; the medium-scale 1-to-62,500 maps cover 15 minutes of latitude and longitude; and the 1-to-250,000 maps cover 1 degree of latitude and 2 of longitude. In the United States 15 minutes, which is a quarter degree, is about 18 miles of latitude and 15 miles of longitude. The latitude and longitude of the map are recorded on the corners and sides. Tick marks on the margin divide the map into nine rectangles, which can be designated for locating a point as in the NW, NC, NE, WC, C, EC, SW, SC, and SE rectangles, or by the position in degrees, minutes, and seconds of latitude and longitude.

All continental land surfaces of the United States except the eastern founding states and Texas are divided into 6-mile squares known as townships. Each township, in turn, is divided into 36 sections, each a mile square containing 640 acres. Townships are numbered north and south from a base line (symbol T) and into ranges (symbol R) east and west from a principal meridian. A township might be Township 3 North (T3N) and Range 2 West (R2W). Sections are numbered from right to left on the odd-numbered lines and left to right on the even-numbered lines, starting in the upper right-hand corner and continuing to the bottom right-hand corner. Sections are divided into quarter sections, primarily for land-title purposes, so that a legal description might be $SW\frac{1}{4}$ of $NE\frac{1}{4}$ of Section 17 of the township described above; its full description would be $SW\frac{1}{4}$ of $NE\frac{1}{4}$, Sec. 17, T3N, R2W. (See accompanying example). By continuing the quartering, an exact location of a single house (or fossil site) any place in the United States where the township system is used can be described in less than one line.

Topographic maps are made today from aerial photographs from which three-dimensional projections can be made to give the mapmaker the data on contours, drainage, forested areas, etc. Maps were formerly made from surveys sketched by hand in the field. Today some such surveys must be made to provide basic points for interpretation of the aerial photographs

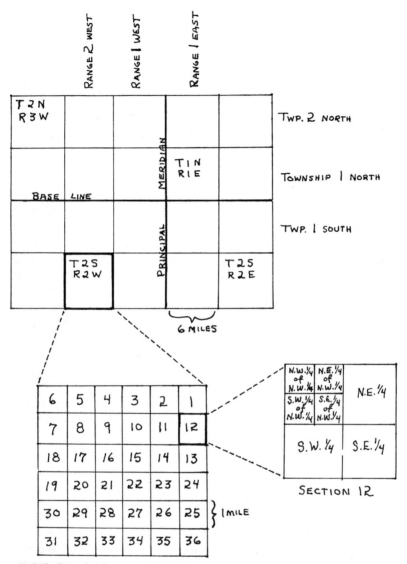

T2S R2W DIVIDED IN SECTIONS

Townships are laid out along a principal meridian and a base line (drawing at top). Ranges are measured east and west from the meridian, townships north and south from the base line. Each township is six miles square. Each township is divided (lower left) into 36 sections, each a mile square, numbered from right to left and from left to right alternately. Each section is then divided into quarter sections (drawing lower center), and each quarter into forties (40 acres) (drawing lower right). In designations T stands for township, R for range, and N E S W for the directions.

and to get information about areas, such as a heavy forest, where the detail would be hidden from the camera's eye. Man-made features are also described from field data.

The user of a topographic map will occasionally wish to determine his own position on it. This he can do by taking compass readings on two visible landmarks that appear on his map. He can then draw in the lines on his map, and the point where they intersect will be on his position.

Hydrographic maps are topographic maps of the bottoms of lakes or seas. They are made from soundings taken with an instrument that projects a signal toward the bottom and then registers the interval of time before it rebounds.

Topographic quadrangle maps for areas east of the Mississippi river may be ordered from the Washington Distribution Section, U.S. Geological Survey, 1200 South Eads Street, Arlington, Virginia 22202. Those for areas west of the Mississippi river may be ordered from the Denver Distribution Section, Geological Survey, Federal Center, Denver Colorado 80225. Indexes of the available maps may be obtained from these centers free of charge. These indexes also list dealers in principal cities who stock the maps. Maps are designated by the names of a town or major natural feature appearing on them. Many other agencies of the federal government issue maps, for example, there are Forest Service maps of national forests. They may be ordered from the Forest Service, Department of Agriculture, Washington, D.C. 20250. Orders should be placed sufficiently far ahead to allow time for the order to be filled and the maps returned.

For students, the Geological Survey has made available a set of twenty-five maps showing typical topographic features, such as mountains, glaciers, and faults.

GEOLOGIC MAPS

Geologic maps are topographic maps with added information of interest to collectors, prospectors, miners, and specialists in such subjects as land use and water supply. A topographic map itself conveys a lot of geologic information to anyone able to read it. Ridges on the map probably stand for bodies of hard rocks, such as dikes, upturned strata, and dense limestone and sandstones; valleys indicate areas of eroded soft rocks, such as shale. Geologic maps, however, carry this type of information much further. They are especially useful to the fossil collector because they will tell him where to find sedimentary rocks, the source of most specimens.

Geologic maps locate and identify specific rock masses that lie on or near the surface with respect to other topographic features. Such maps grew out of the ones that Baron Cuvier and Alexandre Brongniart drew of the Paris basin and those that William Smith made of England in the first two decades of the last century.

The basic geologic laws that these men helped discover still guide map-makers today. These include the law of superposition—that in any pile of undisturbed sedimentary rocks the youngest rocks are at the top, the strata underneath are progressively older, and the oldest are found at the bottom. Another law affirms that water-laid sediments—the happy hunting ground for fossils—continue horizontally in all directions until they thin out at the original shorelines on the edge of the basin. To this belongs the corollary that where a sedimentary stratum disappears abruptly, it has either been removed by erosion or displaced by a fault.

In preparing a geologic map, cartographers assume that similar rocks showing the same succession of strata and not too far separated presumably belong to the same time, and that an unidentified stratum probably can be traced along until it interfingers with a more familiar one that will help in placing it.

Field geologists trench, drill, and sample outcrops to identify the surface rocks and interpret their relationships. A geologic map may stop with surface data, or it may become a study in depth—a geologic column—which is like a vertical cut through the rocks at one or more selected points in the map. Such a detailed map may require elaborate laboratory work. Geologic columns showing the pile of strata and relative thickness are usually printed on the margins of the map. Formations (groups of strata) are the basic units of the geologic map, and each must be extensive enough to be meaningful to the user of the map. Folds, faults, and tilted formations are also indicated. For collecting purposes, the relatively simple geologic map that identifies surface rocks is the most useful.

Dip and Strike

Just as on topographic maps, the means employed to convey the map's message are symbols, textures, and colors. One of the most perplexing symbols on the map, that for dip and strike of rock masses, deserves some explanation because it is commonly used and is especially significant. Many rocks, such as sedimentary ones that were laid down in horizontal strata, have been tilted and otherwise disturbed. They may have been raised into anticlines (hills) or depressed into synclines (valleys), which may later have been eroded.

Dip and strike are more difficult to define than to understand. If a common 3 by 5-inch file card is rested on one long edge on a horizontal surface so that it stands vertically, the compass direction along the top edge of the card is its strike. Strike is a compass direction. Now if the vertical card is tilted to one side the angle of the tilt with the horizontal surface will be the dip. Dip is the inclination of the stratum from the horizontal, measured in degrees, and strike is the direction of the line of intersection along the horizontal plane. The symbol is λ25, in which the longer line gives the compass direction and the short line is the dip, expressed in an accompanying figure in degrees.

On geologic maps, contacts of formations are shown in black lines which are solid if the contact is visible and dashed if it is not. Faults are shown by heavy black lines, and the symbols U or D show the direction of movement of the fault walls. Arrows show horizontal fault movement.

Maps as Major Tools

Maps are major tools of the fossil collector. Just as a woodworker must learn how to use a plane and chisel if he is to produce acceptable objects, so the fossil collector must learn to use his map tools. He must learn to read them and to grasp the wealth of information they hold for an informed eye. The best way to do this is to get a general understanding of symbols and conventions at home and then put this knowledge to the acid test in the field. There the symbols become meaningful. The collector will soon learn to locate areas of suspected fossil-bearing rock on geologic maps and to locate quarries on topographic maps. Areas of closely spaced contour lines mark steep slopes where rock is likely to be exposed. River bluffs can easily be located. The easiest access to an area can be planned by taking a route that goes through areas of widely spaced contour lines.

IX

PREPARING
AND CLEANING FOSSILS

Collecting is only half of the job. Few fossils are found so clean that they are fit to be placed in a collection without further work. Some need only a brushing; some require painstaking treatment to remove rock that obscures the details of the fossil. Proper cleaning is important. Almost every day, an amateur collector brings a fossil to a museum to be identified. Too often a rare specimen has been damaged because its owner brushed varnish on it or destroyed fine detail by plunging the fossil in acid to get rid of matrix.

Preparation begins in the field with use of proper tools. Each person in the collecting party should have a prospector's pick, a flat chisel, and a square-pointed chisel; and the party should share sledges, crowbars, and shovels. Without proper tools, the collector will be unable to remove fossils in an undamaged condition. He will also need a knapsack, a collecting bag, or an apron with pockets where he can stow away wrapped specimens. For some areas, a metal bucket or a basket is more convenient. Putty knives or old table knives are useful for splitting shale, and an old toothbrush will be useful for scrubbing dirty fossils in the nearest puddle or stream.

The shallow cardboard trays that hold four six-packs of beer (commonly called "beer flats" by collectors) can be fitted together if the corners of

Pieces of large turtle shell are cemented together with plaster of paris and dextrin. The plaster fills in chipped edges and can be colored to match the fossil.

the bottom tray are bent in. This makes a shallow, extremely strong, covered box of convenient size to transport or store fossils.

Extremely soft matrix must be treated to harden it for the trip home. Some carbonized fossils, such as plants and fish, must be sprayed to keep the fossils on the matrix, or they will crumble to dust after a few miles of traveling. Vertebrate fossils require elaborate plaster casts before they can safely be moved from their resting places in the field. At home, the actual work of cleaning fossils will begin: clinging matrix can be removed or trimmed to size, and rock can be dissolved or otherwise eliminated to free its content of small fossils.

FIELD WORK

Most fossils found in the field need little care other than wrapping them in paper to prevent abrasive contact with companion specimens on the way home. Loose, sturdy fossils such as brachiopods that are collected from shale exposures can even be piled without wrapping in a tin can or a small box, if the container is packed full so that the contents do not rattle.

A cigar box is excellent for this purpose. When collecting at some sites where there are thousands of loose specimens, this will save much time. Most loose fossils have their own thin protective jacket of mud and shale that acts as a buffer.

A few time-saving hints will expedite packing fossils in the field. Matrix specimens can be wrapped loosely in newspaper taped shut at the ends. Fragile specimens can be wrapped individually in toilet paper by winding it around the specimen in loosely twisted rolls until the specimen is completely bandaged.

A faster way is to layer fragile specimens in a sturdy box or can, separating the layers with sawdust. Any sawmill has mountains of coarse sawdust for the asking. Small amounts accumulate at any lumber yard. The sawdust can be carried in a sack and added to the specimen-collecting box as needed. If the fossils have deep nooks and crannies, grains of the wood have an annoying tendency to lodge there, but they can be picked out.

An emergency method that works well if conditions are favorable is to enclose a fragile fossil in a gob of wet clay or mud. The mudball can then be wrapped in paper and tossed in with the sturdier fossils. This method is especially useful in keeping together the loose parts of a broken fossil. These mudpacks should be removed as soon as possible, before they dry out. Hardened mud becomes difficult to loosen, and the shrinkage during

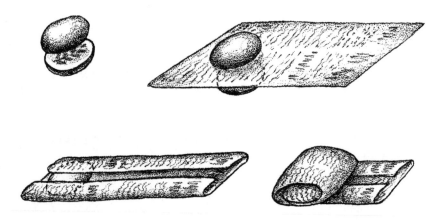

How to wrap a fossil concretion in the field to preserve its fine detail. Place sheet of paper between halves, fold in sides, then roll lengthwise, and secure with rubber band or tape. (Drawing courtesy Illinois Geological Survey)

Broken fossils can be reassembled in aluminum foil. Folded gently over the specimen and pressed tightly, the foil keeps edges from chipping and pieces from becoming lost.

drying may break fragile specimens. Simple soaking at home will remove all the clay or mud.

A fossil found in several broken pieces can be reassembled roughly in a small square of aluminum foil. The foil should be folded over the fossil gently and squeezed to keep the pieces firmly together until the specimen is home. Broken fossils can be mended in the field, but this takes time and often results in a poor job. It is better to protect the pieces and work the puzzle at home. Fast-drying household cements (such as Duco) are suitable for repairing broken fossils.

Don't forget to include a label describing the specific collecting-location in detail. Labels should record the general geographical position of the dig, the assumed geological age of the rock matrix, and the name of the formation and associated formations. Sometimes it is possible to describe the formation by measuring its vertical distance from level ground or a distinctive rock layer. Experience teaches the collector that the best memory is not to be trusted with these technical details, but a good label is forever. Without such a record of the location, a fossil loses most of its cash value and all of its value to science.

Specimens taken from rock layers that are obviously different should be kept separate and should be labeled separately. Loose, weathered specimens taken from the bottom of a slope should be kept separate from those taken directly from an identifiable rock unit. The latter fossils are more valuable because their source is precisely known.

Hardening Matrix

Some shales or weakly cemented sandstones may be so fragile that they cannot be removed without disintegrating. Shales that enclose plant fossils seem to be particularly weak, though fragile invertebrate fossils that must have supporting matrix, such as trilobites, graptolites, and bryozoans, often occur in crumbling shales. These specimens must be hardened on the spot.

A professional concoction invented by the British Museum for protecting and hardening shale containing British Mesozoic fossils consists of two to five tablespoons of flake-form polyvinyl acetate dissolved in a pint of toluene. It may take a day or so for the flakes to dissolve. Polyvinyl acetate is not to be confused with other polyvinyl compounds; it must be the acetate. The mixture is stored in a jar with a tight-fitting lid, since the fumes are irritating and inflammable.

Acetone may be more readily available to amateur fossil collectors than toluene. It will also dissolve polyvinyl acetate. It is inflammable and will cause dizziness if breathed for prolonged periods. Cellulose acetate, in sheets or flakes, can substitute for polyvinyl acetate. These hardeners are similar to such household cements as Duco. In fact, fresh Duco cement dissolved in a few spoonsful of toluene or acetone will serve admirably for small hardening jobs.

The mixture can be brushed on, or the piece of shale can be immersed in the liquid for several seconds. Immersion will make the shale stronger

Shales such as this specimen are poorly consolidated and will turn into mud at the touch of water. Fragile, thin-shelled fossils such as this trilobite will often crumble unless the shale is hardened.

but will also coat the fossil unless the block is hardened by immersing only the backside. The block may need a second dip for super-strengthening. The pieces should be dried in a well-ventilated area; they will dry in a few minutes.

If the mixture contains too much plastic, the fossil will become shiny, particularly if the fossil itself is composed of smooth, nonabsorbent, chitinous material, as trilobites or cephalopods are. If the specimen is too shiny after it has dried, and the fossil can stand rubbing, it can be wiped with a cloth soaked in acetone or fingernail-polish remover (which is perfumed acetone).

These polyvinyl hardeners are also useful for gluing back bits of shelly material that pop loose from a fossil during preparation.

Some fossils, particularly carbonized plant fossils and some Eocene fossil fish of Wyoming, are beautiful when first removed from their stony graves, but as they dry the carbon particles flake off and blow away. What once was a fish skeleton disappears with the drying wind. As soon as the specimen is reasonably dry, it can be sprayed lightly with one of the clear plastic sprays. The specimen should not be sprayed when it is wet; the plastic layer will peel away along with the fossil. Krylon is particularly well suited for spraying fossils as it does not leave a very shiny surface.

Moist shale, which tends to disintegrate as it dries, can be preserved for several weeks if it is sealed inside large plastic bags until there is time to

Loosely attached carbon films, such as fossils of many plants, fish, and graptolites, should be sprayed shortly after collecting with a non-glossy plastic.

work the fossils out of it. This is a method developed in recent years by professional collectors for museums.

CLEANING AT HOME

The carefully unwrapped specimens will need a bath or more extensive cleaning before they can be properly studied or displayed. This is the most tedious part of fossil collecting. Many amateur collectors wonder why specimens in museum are so detailed and sharp, whereas their specimens remain muddy looking. The answer is in the preparation—or lack of it. Museums hire full-time preparators to remove the adhering matrix painstakingly by hand or with machinery.

Some specimens will need nothing more than to be soaked in warm water with a dash of detergent, followed by a scrubbing with an old toothbrush and a rinse of clear water. Specimens that have weathered free from soft shales fall into this category. So do Miocene or Eocene shark teeth and shells found along both coasts in soft, sandy matrix. Concretionary fossils, such as ammonites, bones, crabs, and fern fossils, break to a clean surface and often need no further preparation than washing or brushing to remove dust.

But most fossils, even those that at first glance appear free of matrix, need further cleaning. Brachiopods invariably have matrix wedged in the hinge line. Trilobites seem to have concrete packed in the furrows of their corrugated skeletons. Crinoids have thousands of feathery arms to prepare. Snail openings are obscured with rock. Bony fossils are encased in a rock jacket, but by the time the bones weather free they have become bone meal. In all these cases, hard matrix must be removed.

Soaking and Scrubbing

All hard fossils should first be washed with detergent and water. Hard fossils are durable specimens that are not on a matrix of soft shale or sandstone that is likely to disintegrate when wet, or are not thin delicate films that might loosen in water. Graptolites, carbonized plants, and thin-shelled arthropods are examples of these delicate specimens. When in doubt, experiment with a broken specimen. Some fossils, such as brachiopods found in shale as single shells or valves, are so thin that when the adhering shale is loosened by the water they fall apart.

Small nylon brushes such as toothbrushes are ideal for scrubbing a fossil. Weathered limestone and shales can sometimes be entirely removed

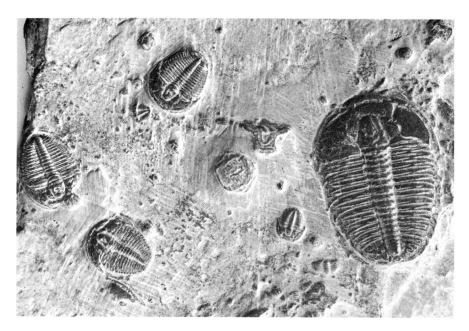

Trilobite *Elrathia kingi* from Wheeler amphitheater, Utah. Specimen in shale was cleaned with a toothbrush and water. Cambrian period.

by gentle but persistent scrubbing. Nylon is softer than the calcite substance of fossils and softer than the matrix, too. It removes only matrix grains that have weathered loose. Hard, fresh limestone and shale will not be touched by brushing. Sometimes soaking for several days in water will soften matrix.

If scrubbing does not remove matrix, set the specimen aside for mechanical preparation.

Stained specimens can sometimes be cleaned by soaking them in a sodium hypochlorite solution (such as Clorox) overnight. Always try this on a sample piece first. If specimens have been permeated with natural crude oil or asphalts (some Silurian fossils of northern Illinois and Indiana are found in this condition), immerse them outdoors for a day in gasoline, scrub them with a brush, then rinse them several times in clean gasoline. Avoid plastic-handled brushes, such as toothbrushes; they soften in gasoline. Allow the specimens to air for a day or two before bringing them inside. Do not pour the dirty gasoline down a sewer: the fumes are explosive. Kerosene or light oils and even the strongest detergents are not as satisfactory as gasoline, because gasoline will penetrate the specimen and remove the crude oil and then will completely evaporate.

Wheeler amphitheater, House range, Millard county, Utah, whose shale beds and limestone strata are highly fossiliferous. (Photo by Dwayne Stone)

Specimens on the surface of soft shales cannot be cleaned in water. The shale will swell and literally explode. These shales are best hardened from the back with the hardener mentioned above, and the fossil itself can be cleaned by gently wiping it with a cloth or paper towel soaked in alcohol. Try to keep the alcohol from soaking into the matrix. When the specimen is cleaned, an allover brushing with the hardener will protect the surface and waterproof the specimen.

Removing Matrix

After preliminary cleaning of the fossil, bits of matrix may still mask important details. Removal of these pieces to aid identification and create a more attractive display specimen requires skill and infinite patience. It takes at least an hour to clean a small trilobite which has soft, relatively easy-to-remove shale embedded between its ribs. It may take days to clean a limestone slab with many crinoid crowns.

The principle behind cleaning matrix from a fossil is to exploit the difference in hardness between fossil and matrix, or, if acid is used for cleaning, the difference in chemical composition. Even though a fossil replaced by calcite is embedded in limestone (also calcite), there seems to be a slight difference in hardness between the fossil and matrix, particularly after some weathering. The calcite grains of matrix, moreover, do not have a close bond with the calcite crystals of the fossil, so that there is a plane of weakness between fossil and matrix.

Softer matrix can be brushed or scraped away or flaked off along the plane of weakness. This is easy if the fossil is composed of a very hard substance (such as pyrite or quartz) and the matrix is something soft (such as shale). It is more difficult when calcified fossils are embedded in limestone. It is nearly impossible when a soft fossil (such as a bone) is in hard matrix (such as limestone).

Exposing Fossils with Hand Tools

The amateur collector is likely to have only hand tools available for cleaning fossils. Only a few types of fossils cannot be satisfactorily cleaned with hand tools. Power tools and such specialty tools as the airbrasive machines are used primarily to speed up the task.

Most specimens will not need major surgery—only a little picking, scraping, and brushing to clean the fossil. However, some pieces will need heavy work before they can be prepared with small instruments.

MAJOR SURGERY

There is a temptation to trim off as much matrix as possible in the field to save weight. Especially with limestone slabs, this can result in breaking the fossil in half where the fossil creates a weak spot in the slab. It is safer to take the heavy slab home and trim and saw it to size there.

If the block enclosing the fossil is too large to transport, a channel should be cut around the fossils with a small cold chisel and hammer. The chisel should be held so that it points away from the fossil, and the channel should be a reasonable distance away from the fossil. After much labor, the fossil will be atop a pedestal, which can then be broken from the block with a single blow from the hammer against the chisel placed at the base of the pedestal. The deeper the channel and the higher the pedestal, the better the chance of removing the fossil in one piece.

If the block, however unwieldy, can be hauled home, the chances of removing the fossil safely are much better. Limestones, hard shales, slates, and sandstones can be sawed to convenient sizes and shapes with a diamond saw blade such as is used in lapidary work. Small slabs can be cut

Diamond-charged saw blade, using water as a coolant, trims a fossil block.

quickly into square pieces, and individual fossils can be separated with little or no damage with a small trim saw using a six-inch to ten-inch blade. Such trim saws can be purchased new for $25 to $50, not including blade and motor. A six-inch blade costs less than $15 and a ten-inch blade about $25. If used only for sawing sedimentary rock, the blade will last the fossil collector many years. The saw coolant should be either a water-soluble oil or one of the materials marketed to be added to tap water. Water alone will rust the blade.

For lapidary work, in which rocks much harder than those usually encountered in fossil work must be sawed, a light cutting oil or kerosene is used as a coolant. If sawing must be done on a saw that uses oil as a coolant, the specimen should be soaked in water for several hours prior

Removing a fossil from a block of limestone. Step 1: First chisel a groove about an inch from the specimen. Tilt the chisel away from the specimen to avoid chipping it.

Step 2: After the groove is 3/4 inch deep (or more if the fossil is large) place the chisel at the bottom of the groove and angle it sharply toward the fossil. A sharp blow with a hammer on the chisel should then break loose the fossil atop its pedestal.

Step 3: The result. The fossil can be trimmed further at home if desired.

to sawing. After sawing it should be plunged into a strong detergent solution. Soaking will prevent too much oil from penetrating porous limestones or shales. If matrix will not disintegrate, the specimen should soak for several hours or days in the detergent.

Specimens too large for a trim saw may require commercial sawing, a service usually available at a rock shop at a cost of about ten cents a square inch.

Thin slabs of shale or limestone can be sawed with a silicon carbide cutoff disc on an electric drill or flexible shaft machine. These cutoff discs are not expensive; but they create annoying dust, and they are fragile. A steady hand is needed to keep them from breaking.

Cement and tile contractors use a special saw for trimming tile or marble slabs or for sawing concrete. This outfit uses a much thicker blade than a lapidary saw, usually with a water coolant. The saw blade is overhead and is lowered onto the work by a foot-operated lever. The blade is usually a diamond blade, but it may be silicon carbide, which will easily cut limestones or shales. Such a saw is excellent for trimming large pieces.

Carbide hacksaw blades and diamond-impregnated wire blades that fit in a coping-saw frame are also useful for sawing matrix.

If no saw is available, the waste material can be removed carefully with

A silicon carbide cutoff disc mounted on a flexible shaft is useful for trimming small fossils.

hammer and chisel. A small channel should be chiseled across the matrix where it should break. The slab can then be placed on a flat block, a curb, a concrete step, or some other hard surface with a square edge. The part to be broken away should hang over with the chiseled line facing up and at the edge of the support block. While the piece is held firmly, the protruding end should be given a sharp blow with a heavy hammer. If all goes well, the piece will snap cleanly along the line.

Small edges can be chipped off, using the chisel tilted slightly away from the fossil. With solid, fine-grained limestone that breaks with a conchoidal (shell-like) fracture, the edge can be chipped away quite rapidly with sharp blows from a small hammer. If a piece is thin, the edge can be nibbled away with pliers. Small bites, not more than a fraction of an inch, should be taken, then the pliers should be tightened and snapped down or up sharply.

If only a small amount of material is to be removed, it can be ground

Orville Gilpin, chief preparator at the Field Museum, does most of his work with a small hammer and a series of chisels. Fine work is done under the binocular microscope. The sand bag cushions the specimen and reduces noise.

Bits of the shell of a fossil turtle are scattered through a block of sandstone. Orville Gilpin chisels them out one by one and cements them to the reconstructed piece.

away on lapidary-type silicon carbide grinding wheels. Small high-speed wheels used for sharpening tools are not well suited for grinding fossil specimens. Pieces weighing more than a few ounces should never be ground on a wheel, as they will quickly create a bumpy wheel surface or even cause the wheel to shatter, shooting out pieces of sharp grinding wheel at the speed of a bullet. Grinding should be done on an individual fossil only to remove unsightly matrix or to flatten a side or back so that the specimen will sit properly when displayed.

Large pieces can be broken cleanly with a hydraulic press using a hardened steel or carbide point. Grooving the piece with a chisel will help to make a long, clean break, or the specimen can be trimmed more safely by biting off small corners, one at a time. Such hydraulic units are expensive.

Some soft shales cannot be sawed because the water or oil coolant will disintegrate the shale. Such specimens can be trimmed by nipping with pliers, breaking with a hydraulic press, or sawing dry with an old hacksaw blade. The hacksaw blade will wear quickly, but as long as some teeth remain it will continue to cut.

When the fossil is a convenient size to handle, the delicate task of removing the remaining matrix can be begun.

Trimming slabs of stone by nibbling at the edges with pliers.

FINISHING TOUCHES WITH HAND TOOLS

Fossils that are harder than their matrices are the easiest to clean and luckily the most common. Pyritized, calcified, or silicified fossils in weathered limestone or shales soft enough to be scratched with a finger-nail fall into this class. So do loose fossils weathered from shales or chalky, impure limestones.

A vigorous brushing of adhering matrix with an old toothbrush and water will show whether the matrix can be removed this way. Such brush-ing will not damage the surface of any sturdy fossil (brachiopods, most trilobites, corals, blastoids, crinoid cups, etc.), but should be done with care on lacy bryozoans, graptolites, crinoid crowns, and the like. If the matrix is not too thick, a surprising amount of cleaning can be done in a short time. This is the only way to clean out the fine furrows on the surface of a brachiopod. Soaking a specimen for several days in water may help loosen matrix.

If the fossil is considerably harder than the matrix, the brushing can be done with a fine-bristled brass brush, such as a suede brush. On pyritized, silicified, and even some calcified specimens, brass brushing will remove matrix more rapidly than a toothbrush and will remove some hard shales untouched by a nylon brush. Pyritized Devonian starfish and crinoids from Bundenbach, Germany, can be cleaned this way, as the pyritized fossils are encased in a tough shale. Devonian trilobites from western New York and Ohio (generally black or dark-colored and preserved in a gray shale) can be quickly cleaned by gentle use of a brass brush. Pyritized brachiopods from Ohio respond equally well.

It is easy to tell whether the fossil is harder than the brass brush and therefore is not likely to be damaged. Brush a specimen, and if the fossil quickly picks up a brassy shine it is harder than the brush. If the matrix wears away and does not turn a brass color, it is softer. The brassy color can be removed from the fossil by scrubbing with a nylon toothbrush and a detergent.

Fine steel-wire brushes can be used, but only on silicified or pyritized specimens. Try a sample first to see whether the brush wears away the fossil. Both steel and brass brushes are made for use on a flexible-shaft machine.

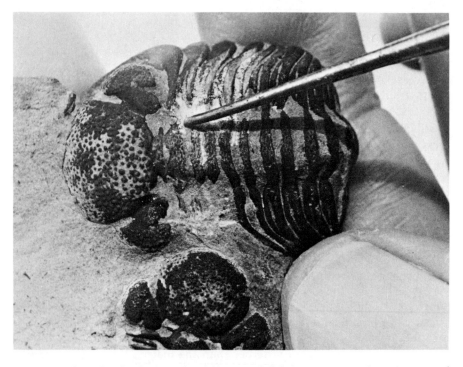

A sharpened crochet hook is used to scrape and flake away matrix from a pyritized trilobite. The entire specimen can be prepared this way and finished by brushing.

Experimentation on a corner showed that the trilobite was harder than a brass bristle brush and the matrix was softer. After most of the matrix was removed with the scraper, a small brass wire brush on the flexible shaft completed preparation of the specimen.

Most specimens will have some areas of thick matrix that can be removed more rapidly with metal scrapers. Small-bladed pocketknives make fine scrapers for large areas. Screwdrivers can be sharpened to various shapes for scraping or prying loose larger pieces of matrix. For delicate work, fine, needlelike scrapers should be made. Discarded dental tools fit the hand and are usually available from your dentist. Crochet needles work well if the hooked point is ground away. A handle can be made for them from a piece of wooden dowel drilled to accept the shaft. If one can still be found, a steel phonograph needle is a particularly tough, fine point for delicate work. Ordinary needles, mounted in a wooden dowel, work well. Large needles used for sewing carpets can be used for large-scale work. Points must be kept sharp, as the matrix quickly dulls them. A dull tool may slip and damage the surface of the fossil.

The scraping action should be much like cleaning fingernails, with just enough pressure to move the matrix but not enough to bite into the fossil.

The trilobite, completely prepared with a scraper, brass brush, and final scrubbing with a toothbrush to remove the brassy shine.

Brush and wash frequently. Don't try to scrape off that last thousandth of an inch. With practice, little damage will be done to the fossil.

A trickier technique is flaking away adhering matrix. This works particularly well on fossils that are smooth and shiny and do not have many indentations. It also work well on some shale-encased fossils, such as the Wyoming fish or the plant fossils of the Mazon Creek area in Illinois. The trick here is to start a few millimeters away from some exposed part of the fossil, dig the pointed scraper into the matrix, and with a sharp push and twist flake away a piece of the matrix. It should part cleanly from the fossil. This can also be done using the scraper and a tiny hammer, angling the tool toward the exposed fossil at about 45 degrees. A single blow should dislodge the flake. Never try to slide the scraper between the matrix and the exposed fossil. The matrix will flake away, but a splendid scratch will remain on the fossil. This flaking technique can be done rapidly in a routine of dig in, flick upwards, and blow away the chip.

Flaking works best when the matrix is thin. This may require a bit of preliminary grinding, a rather dangerous adventure that can also destroy parts of the fossil that protrude unexpectedly.

Power Tools

With the exception of a sandblaster machine, power tools are only labor- and time-saving versions of hand tools. They are easier to use, but the power-operated instrument grinds relentlessly on, ignoring the interface between matrix and fossil. Mistakes are easier to make, and by the time the machine has been stopped or the power head moved the damage has been done. Dust and matrix chips obscure the work so that it is easy to go too far.

VIBRO-TOOL
Several types of vibrating-point engravers are available; the one most commonly used is the Vibro-Tool. All are priced around $10.

The vibrator in such a tool causes a sharpened point to strike tiny blows many times a second, rather like a miniature jackhammer. An adjusting-screw controls the length of the stroke. The best model for fossil work has a chuck that will hold different points.

Vibrating tools are useful to remove matrix rapidly from such specimens as these fossil clams.

The steel points usually provided with vibrating tools are too large and too soft for hard stone, but they are serviceable, particularly if kept sharp and not used too often. For fine work a sharpened dental burr or a phonograph needle is excellent, but since these points will not fit into most chucks, an adaptor must be made of a short piece of brass rod that does fit the chuck. The brass rod should be bored lengthwise for half an inch to accept the phonograph needle or dental burr. Another hole should be drilled at right angles to the first one and tapped to accept a small setscrew, which will hold the small point tightly in the hole. The unit can then be placed in the chuck of the tool and tightened securely.

As the tool hammers away it kicks up a cloud of dust, hiding the work and fouling the air. The dust can be blown or whisked away, which causes constant starting and stopping of the tool, or the work can be done outdoors in a strong wind. A small air hose fastened to the tool and connected to a strong pair of lungs or an air compressor will blow away the dust as it forms. A section of thin plastic tubing can be taped to the side of the tool with its opening near the working end, so that the operator can blow through the other end of the hose.

An inexpensive source of compressed air can be obtained from a small aquarium aerator. A more sturdy arrangement can be made with copper or aluminum tubing carrying the air across the Vibro-Tool from the plastic air hose at the top of the tool. A compressed-air blower system based on the aquarium aerator can also be attached to a stand and aimed at work being done with hand tools.

The action of the vibrating tool can be compared to that of a tiny hammer and chisel. The rapid hammering of the pointed tip of the tool flakes away the matrix. The best action seems to occur when the tool is held almost at right angles to the specimen. A short stroke is used for fine work; the long stroke bangs off larger pieces of matrix and is good for removing large areas of unwanted stone where there is no danger of breaking through into the fossil. As the fossil surface comes near, only fine vibration and short strokes should be used. Usually small pieces will break free at the fossil-matrix interface shortly before the point digs into the fossil surface.

Hours of patient work and a thorough knowledge of the morphology of the underlying fossil are necessary before you will know when to stop. If the fossil is likely to be distorted or bent or to bear unsuspected projec-- tions, work should be done by hand, at least until the general topography is revealed. The vibrating tool is too fast and undiscriminating for blind work.

Allowing the tool to work straight down in one place usually will loosen a chip of matrix in a manner similar to the flaking method used with hand tools. If the tool is moved back and forth smoothly the action is rather like that of a hand scraper. Different matrices require different techniques; each specimen is different.

The tool will become uncomfortably warm if used for a long time. If the vibrations carried through the stone reverbrate like a drum from the table below, an effective damper can be made from a small cloth bag filled with sand. The work is laid on the sandbag. The sandbag molds itself to the contours of the matrix and helps hold the specimen while it is being prepared. Sandbags are useful for any type of preparation.

The Vibro-Tool is sold by Burgess Vibrocrafters, Inc., Grayslake, Illinois 60030. Other vibrating engravers are available at most craft and hobby stores.

POWER DRILLS

Dentists have found that one of their best tools is a high-speed power drill that will quickly and cleanly remove unwanted tooth material. Such a drill is equally effective with fossils. Unwanted matrix disappears quickly before a burr, tiny grinding wheel, or brush mounted in the chuck of a dental drill.

A used dental drill can sometimes be purchased for a moderate sum, but never cheaply. Such a unit hanging above the workbench is a fine addition to fossil-cleaning tools.

For a smaller sum a small unit operating through a flexible shaft will give almost as much versatility and movement as the dental drill. These units usually consist of a motor the size of a soup-can that is hung above the workbench. A flexible shaft several feet long hangs from the motor and ends in a chuck that will accept a wide range of commercial and home-made abrasive tools. These units can be purchased new for about $35 and are handy accessories in jewelry making and fine metal work. The flexible-shaft hobby machines allow absolute control of speed through a foot-operated rheostat. So do the dental drills.

Tiny hardened steel burrs mounted in the chuck will tear away excess matrix and, if handled carefully, will not materially damage fossil parts they might brush against. Tiny silicon carbide grinding wheels or cut-off discs can be used to remove large quantities of excess matrix to improve the appearance of a specimen or to reduce the thickness of matrix overlying the fossil preparatory to use of some other cleaning method.

With a flexible-shaft tool, small brass and steel brushes are particularly useful. These brushes, about one inch in diameter and less than a one-quarter inch thick, will scarcely touch hard fossils, especially pyritized or silicified ones, but will rapidly remove hard shales or limestones, as has been mentioned in the section on cleaning with hand tools.

The brass brush should not be used for soft fossils after it wears to less than half its original width. The bristles are stiffer then and are more likely to cut into the fossil than to bend while going over it.

Bristle and nylon brushes are also available for the flexible shafts. They do what a hand-operated toothbrush can do but in a tenth of the time.

Burrs come attached to a small-diameter shaft. The wheels and brushes can be purchased attached to a shaft or loose for mounting on a threaded shaft. The loose tools are usually less expensive. When using them, it is rarely advisable or necessary to run the flexible-shaft machine at full speed. Low speed allows better cutting, as the bristles dig into the matrix instead of sliding over it, and, of course, control is easier. Too much pressure strains the flexible cable, which is expensive to replace. Too much pressure also bends the small shafts of the tools.

Most fossils can be completely prepared using grinding wheels or burrs to remove excess matrix and brushes to remove the last vestiges of matrix. Good control can be had with power equipment, particularly with power brushes, and the greatly increased speed reduces mistakes that arise from fatigue and impatience with the slow progress with hand tools.

SANDBLASTING FOSSILS

The sandblasting process, used to clean buildings and to etch tombstones, can also be used to clean fossils. A few years ago a small sandblaster that allowed localized and precise cutting was put on the market by Pennwalt-S.S. White Dental Products Division, 3 Parkway, Philadelphia, Pennsylvania 19102. It was designed to replace the dentists' drill. The idea was excellent, but the machine proved impractical because the abrasive clogged the fine orifice. Similar units are now available from the same company for etching metals, glass, and plastic and are easily adapted for use in cleaning fossils.

Almost every major university and museum in the United States now has one of these machines, and some private collectors have them, too. They are the ultimate in fossil cleaners. The abrasive stream can be directed so that layers of fossils are revealed without removing them from the matrix block. On crinoid slabs from Le Grand, Iowa, several feathery crinoids may lie atop one another. Old preparation methods could prepare the top layer well and perhaps part of the underlying one, but too much work on the bottom layers would break away parts of the top. With the gentle action of the sandblaster, all parts can be undercut and cleaned perfectly in areas where the tiniest brush couldn't reach.

Unfortunately, there are some drawbacks: a sandblaster costs several hundred dollars, the abrasives are somewhat expensive, and the machine is rather delicate. Dried compressed air or carbon dioxide must be used

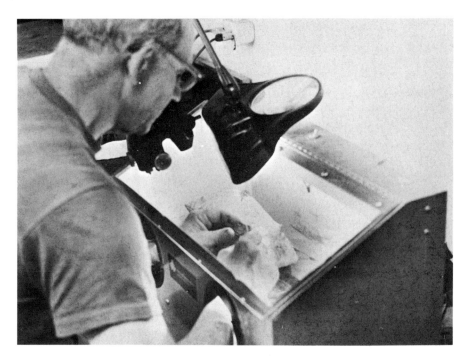

Using the abrasive machine to prepare a delicate fossil. Work must be done within a tightly closed box.

because the machine will clog if there is any humidity in the air. The type of fossil and matrix must be well understood—use of the wrong abrasive will destroy the fossil.

Basically, the machine consists of a vibrating hopper filled with fine abrasive which is fed into a stream of rapidly moving gas. The abrasive particles shoot out of a tungsten-carbide nozzle, bounce off the specimen, and wear away the material. The actual speed of the particles approaches 1,100 feet a second—about the speed of sound—and at this speed the abrasive will cut rapidly even though it is of the same hardness and composition as the matrix being removed.

The speed of cutting is controlled by the amount of abrasive fed into the stream of gas. The nozzle comes with a round hole about 1/50 inch in diameter for general work or with a narrow rectangular hole for special work such as precision cutting. The hand piece is jointed so that the nozzle can be aimed in any direction for working in difficult areas.

Coarse removal of matrix is done with aluminum oxide of about 50-micron diameter, and finer work with a finer size of the same material.

Powdered dolomite (calcium magnesium carbonate) is most commonly used for general work when not too much matrix must be removed from calcified fossils. For exceedingly delicate specimens, such as paper-thin shells, trilobites, and teeth, powdered sodium bicarbonate is used.

Work is done in an airtight box fitted with arm holes that have tight rubber sleeves. The top has a window of glass made to be replaced easily as it will become frosted through use of the machine.

Ultrasonic Cleaners

Weathered matrix and clay often remain firmly wedged in fine openings in fossils, particularly such pore-bearing fossils as bryozoans and corals and around the compound eyes of trilobites. All the brushing in the world won't remove this material. But just a few minutes in an ultrasonic cleaner will remove this debris, and even some that appears firmly attached. Sand and shale that are not too firmly cemented can be removed neatly by placing the specimen in an ultrasonic cleaner tank for fifteen minutes or so. These cleaners are also used to clean delicate or intricate parts, such as the works of watches. The sonic bombardment loosens dust or dirt.

The ultrasonic cleaner consists of a tank to hold the solvent in which an object is suspended and a source of inaudible, extremely high-pitched sound waves. Bombardment by these sound waves of the liquid in the tank produces cavitation, which is rapid formation of vapor pockets and their collapse. This causes a violent, small-scale, hammering action on an object suspended in the liquid. Anything not firmly attached is torn loose.

There are two basic types on the market: a high-frequency, high-energy unit that is heavy, expensive, and rather dangerous to operate, and a low-frequency unit that is portable, not so expensive, and safe to use. The low-frequency units seem to do a satisfactory cleaning job, and the magneto-strictive transducers that produce the ultrasonics will not overheat and can be used for a long time. Small units with a container large enough to handle a specimen several inches in size are now on the market for around $60. The price increases rapidly for larger units.

While most commercial units distribute the ultrasonics throughout the tank, one has been built with the vibrations coming from the end of a rod, which can be aimed at certain parts of a submerged fossil specimen to give particularly vigorous cleaning action in a localized area. Such a directional ultrasonic cleaner was designed by English paleontologists and displayed in England at a meeting in 1967. This unit did a superlative job of cleaning sandy matrix from fossil bones. A tank-type ultrasonic cleaner used long enough to disintegrate the sandstone would very probably disintegrate the bone as well, but a directional unit could be aimed at the matrix without undue damage to the bones. Such directional probe type ultrasonic cleaners are now being marketed in the United States.

Major manufacturers have announced that they will be producing an ultrasonic cleaner for use in the kitchen sink. The vibrations that will tear loose congealed steak grease may find application in the fossil-cleaning field as well. A cleaner such as this would loosen all but the most stubborn matrix—affording a quick, easy way of preparing fossils for the final detailed work by hand or with other machines.

A typical cleaner, made by the Bendix Company, has a power output of 180 watts and produces sound from 19,000 to 22,000 cycles a second. The cleaning receptacle has a capacity of $1\frac{1}{4}$ quarts, more than sufficient for most fossils.

In practice, the specimen is placed in a beaker which is then placed in the tank. Water with detergent or wetting solution is an excellent liquid for cleaning the fossils. With delicate microfossils, immersion time is in the range of one to two minutes. With typical bryozoans, brachiopods, corals, and echinoids, the time may be increased to ten or fifteen minutes. By then most of the material that will come loose has done so. Further treatment may break apart the fossil.

Generally, little damage will be done to the fossil, though long treatment will create some surface abrasion. Delicate appendages, such as brachiopod spines, may be broken off, especially if the power source is strong.

Specimens being treated with acid, or limestone blocks being dissolved to free enclosed silicified fossils, can be placed in the machine in beakers of acid. Short periodic bursts of ultrasonic vibrations will materially speed up the solvent action of the acid.

The ultrasonic cleaner cleans mineral specimens as well, particularly dirty geodes or sturdy crystal groups with clay trapped in crevices. A short blast of the machine will clean embedded grit from pieces of tumbled agate and will remove polishing powder from cavities in slabs or cabochons.

Never add grit to an ultrasonic cleaning solution in expectation of faster cleaning; it will quickly destroy the specimen.

Boiling

For years, boiling shale with chemicals has been a favored way to release its hidden fossils. Varsol or trisodium phosphate dissolved in water has been used for this purpose. Alternate wetting and drying, using water alone, does the same thing. Soaking in gasoline will soften some shales rapidly.

QUATERNARY-O

A chemical has recently appeared on the market that is far better than any of those formerly used. The product is Quaternary-O, which in the words of the manufacturer is a "high molecular weight quaternary ammonium surface active agent." Translated, this is a super-detergent with superior wetting action. It is a thick, brownish goo resembling automobile lubrication grease. It dissolves slowly in hot water, but not in cold. The fossils are placed in a saucepan and covered with a solution of Quaternary-O. Several tablespoons of Quaternary-O dissolved in a pint of water is a strong enough solution. The liquid containing the fossils is brought to a rolling boil, and boiling for ten to fifteen minutes cleans most specimens. Longer boiling, up to half an hour for stubborn specimens, does not seem to hurt the fossil despite the slight acidity of the solution. As the liquid boils down, more water should be added.

After the liquid cools, the fossils are ready to be removed. The clear liquid remaining after the solution stands awhile can be poured off and reused almost indefinitely. There is a slight odor during the boiling, but it is not unpleasant. Do not plunge hot fossils into cold water; they might shatter.

Fossils that seem particularly well suited to Quaternary-O cleaning are those from shale, sandstone, or somewhat weathered limestone matrices that are not too thick. Even rather thick matrix can be removed layer by layer if the boiling is combined with scraping to remove the layer of loosened material before the specimen is boiled again. Any fossil that has weathered free of matrix is usually sturdy and will clean well.

The boiling is almost as effective as ultrasonic cleaning in its ability to remove matrix from intricate pores and openings, and to reveal borings in brachiopod shells, pores in blastoids, and surface ornamentation in clams and brachiopods. Such details may not have been visible even after scrupulous cleaning by hand. As with all types of cleaning, a good practice is to try a damaged specimen first to see whether there is any reaction to the Quaternary-O and whether the fossil will withstand bouncing about in the pan under a rolling boil.

The sludge removed from the bottom of the boiling pan can be allowed to settle through several changes of water in a jar. It will contain perfectly cleaned microfossils such as ostracods, conodonts, microcrinoids, fish teeth and bones, and who-knows-what. If fossils from only one locality are boiled at one time, and the sludges are kept separate, the microfossils will not be mixed. Thus they will be a bonus along with well-cleaned fossils.

Quaternary-O is available from Ciba-Geigy Chemical Company, Ardsley, New York 10502. It costs roughly $1 a pound. Ask about packaging when writing the company for prices, and try to avoid having small quantities sent in glass jars which require expensive special handling in the mails.

Acids

Nature's way of cleaning fossils is with acid, but she has more patience than the fossil collector. A gentle wash with carbonic acid so weak that we drink it daily without realizing it will clean fossils locked in a carbonate matrix. This is how nature does it, and it takes a few years. We can dip a fossil in dilute hydrochloric or acetic acid and do in ten seconds what nature would take ten years to do. But we also run the risk of ruining the fossil in the attempt.

Carbonic acid is carbon dioxide dissolved in water. It forms naturally in bubbling hot springs and in rainfall as the drops absorb carbon dioxide from the air. Ground water also picks up carbon dioxide released from decaying and living organic material. When any of these carbonated waters touch limestone, they slowly dissolve it. For some reason the calcium carbonate of the limestone is dissolved much more readily than the calcium carbonate of a fossil brachiopod.

All carbonated beverages from beer to cola are acid enough to dissolve rock. A brachiopod plunged into a can of cola will be cleaned if it is allowed to remain long enough in the can and if the carbonation is not allowed to escape. However, it is better to drink the beverage and clean the fossil some other way.

Stronger acids, when used properly, can expose and loosen fossils; but when used on the wrong types of fossils or in the wrong way, they can destroy them. A fossil should never be treated with even the weakest acid without first trying the acid on a broken specimen of the same type. Let the broken piece dry, then examine it, for much damage can be masked by a film of water.

Three types of acids are used in cleaning fossils. One type dissolves a matrix consisting of quartz, such as sandstone, without harming calcified fossils that are in it, such as brachiopods or crinoids. The second type includes the weak organic acids used for gently removing carbonate matrix from carbonate fossils. The third type has the strong acids, used for removal of carbonate matrix from silicified, pyritized, bony, or tough carbonate fossils.

ACIDS: TYPE 1

The first type consists of only one acid, hydrofluoric, which is used to dissolve quartz. The amateur should not use it unless he has adequate facilities and a background in chemistry. Hydrofluoric acid is a close relative of hydrochloric acid, but hydrofluoric acid is the meanest member of the family. A finger can be dipped in strong hydrochloric acid without too much damage, but dip it in hydrofluoric acid, and in a few days ulcers will develop that cause deep scarring. Hydrofluoric acid dissolves quartz; it will dissolve a glass bottle. It corrodes most metals. The fumes alone are sufficient to cause skin damage. It must be used in plastic containers in a well-ventilated chamber, such as under a fume hood in a chemical laboratory. It is not an acid for kitchen or basement use.

Actually, there are few occasions when fossils must be removed from a silicate environment. Fossils found in sandstone are usually poorly preserved and hardly worth the danger of acid burn. Occasionally, well-preserved calcified fossils will occur in sandstone. Such fossils can be successfully prepared with hydrofluoric acid. One such recent occurrence was a pocket of complete blastoids found in a sandstone channel-fill in early Mississippian rocks of Montana. The small pocket yielded hundreds of blastoids complete with stems and brachioles.

Since the calcified fossils were much softer than the matrix, any mechanical preparation would have destroyed them. But excellent results were achieved by placing the specimens upside down in a wide, flat, polyethylene dish under an acid hood. Technical grade (48 percent) hydrofluoric acid was poured in to a depth of $\frac{1}{4}$ inch to eat away the fossiliferous side of the blocks. After thirty to forty-five minutes the slabs were removed from the acid with platinum-tipped tongs held with rubber-gloved hands. The slabs were then placed under running water for a minute or so. If necessary, the specimens were etched again by further acid treatment. The fossils needed only a little soap and water and brushing to complete the cleaning.

Hydrofluoric acid has one other use. Small fossils and microfossils show internal features if they can be made translucent. This can be done by changing their chemical composition from calcium carbonate (calcite or aragonite) to calcium fluoride (fluorite). Calcium carbonate will react with dilute hydrofluoric acid to do just this chemical magic, and the fossil will become a pseudomorph of itself without losing a single dimple. It will become translucent when wet and more stable chemically as well as a bit harder. The hardness of fluorite (4 on Mohs' scale) is higher than that of calcite (3 on the scale). This technique works quite well with small brachiopods, ostracods, foraminifera, and bryozoans.

Hydrofluoric acid is expensive as well as destructive. It should be stored where it is not apt to be spilled and where it will not cause major damage if it is spilled.

ACIDS: TYPE 2

The gentle organic acids—acetic and formic—are used to clean fossils by slowly dissolving calcite (limestone). Acetic acid gives vinegar its tart taste and smell. White vinegar is diluted acetic acid, formed naturally by fermentation of alcohol. Formic acid, which causes the excruciating sting of an ant bite, is also found in some stinging plants.

The two acids are quite similar in action but not in price. Acetic acid is far cheaper. Formic acid may be a bit gentler on some fossils, but not enough to justify the difference in cost.

Plain white vinegar is a satisfactory, though slow-acting, source of acetic acid. Glacial acetic acid, a strong concentration, can be bought at a photo-supply store or chemical-supply house. Glacial acetic acid should be cut with five to ten parts of water to make a solution usable on fossils.

After a fossil has been thoroughly prepared, a thin layer of limestone or limy shale may remain. Mechanical removal would undoubtedly damage the fossil. This layer should be as thin as possible before acid removal is considered, and the acid should be the last resort. The specimen is placed in the weak acid in a glass or plastic dish (not metal) and left for a few seconds to observe progress of the reaction. The acid will eat through the matrix and will also eat the fossil, though much more slowly.

Fossils prepared with acid have an unnatural, polished appearance which may be objectionable. Some detail will be lost, and this may counterbalance any benefit from removing the last vestiges of matrix. The acid may be placed with a small brush on spots of matrix, little by little, until the spot

A pyritized brachiopod *Mucrospirifer* looked like this before acid treatment. Matrix is shale, but contains enough calcium carbonate to allow disintegration with acid.

After immersion for one minute in hydrochloric acid, all matrix is gone, but so is a good bit of the fossil detail. The shiny, brilliant pyrite is of little value except as a curiosity.

disappears. When bubbling stops, the spot should be brushed to remove loosened stone and to check progress. This is a slow method but it permits exact control of the preparation.

Some delicate silicified fossils can be removed from carbonate rocks— for instance, Miocene insects from the small California nodules—by the use of acetic acid. These nodules take weeks to dissolve in a plastic container of acetic acid, but the gentle action leaves the insects intact. Such delicate features as legs and antennae are undamaged.

Acetic acid will not dissolve fossils composed of calcium phosphate, such as conodonts and a few species of brachiopods. These can be removed from carbonate matrix by submerging the blocks of limestone in vats of acid until they are dissolved. The mud at the bottom of the tank can be washed away gently with many changes of water. The remaining fossils when dried will be ready for mounting.

To test a fossil to see whether it is silicified or has been replaced by calcite, put a drop of strong acetic or hydrochloric acid on it and watch through a magnifying glass for signs of bubbling or fizzing. If the acid causes fizzing, the fossil is calcite.

Some weathered fossils will be coated with white calcium carbonate

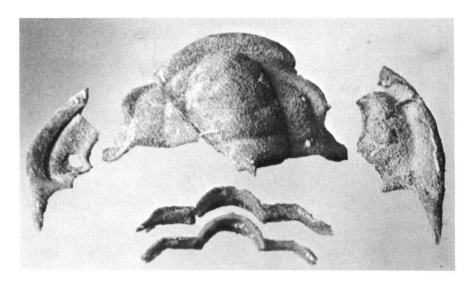

Fragile pieces of *Yorkella australis*, a South Australian silicified trilobite, etched from limestone with acetic acid.

deposited by ground water. A short dip in weak acetic acid will remove this coating without harming the fossil. This is commonly done with concretionary fern fossils from Illinois or Indiana that are found exposed as a result of weathering. Invariably the plant fossil is obscured by a white mask, which is probably derived from a chemical reaction between the concretion and water. After acid treatment, the loosened film will brush away with soap and water.

ACIDS: TYPE 3

Three other strong, cheap, and easily obtained acids are contained in this group—hydrochloric, sulfuric, and nitric. The last two are dangerous to use and have no advantage in fossil use. Hydrochloric acid, also called muriatic acid, is an excellent solvent for any carbonate matrix, acting faster than organic acids though not as gently.

Hydrochloric acid should be diluted to about a 10 percent solution, or weaker, for fine work. Cold water should be used, because mixing acid and water creates enough heat to crack a glass dish. Always add the acid to the water. Remember the triple A: Always Add Acid. Acid is heavier than water and when poured into water will sink, mixing as it goes. It should be stored in a container made of glass, stone, or plastic—never one made of metal—and the tightly closed bottle should be kept away from metal objects. Vapors will escape from the tightest bottle and will rust nearby metal. Fumes from an opened bottle of acid are irritating, so the pouring should be done outdoors while standing upwind. Fossils placed in normal dilutions of acid can be fished out with the bare fingers; gloves or tongs are not necessary unless there are unhealed cuts on the fingers.

Hydrochloric acid is used to free fossils composed of quartz and pyrite. It is also used to free rare anhydrite fossils. Fossils replaced by silica, such as quartz, opal, agate, jasper, chert, or flint, are not uncommon, especially in some areas. If some fossils in a rock layer are silicified, most of them will usually be silicified. Some stone blocks contain enough calcium carbonate to cause them to dissolve or fall apart when treated with acid, and their delicately preserved fossils can be released in this way.

Natural weathering of limestones and limy shales that contain silica fossils will free the fossils, but it may take years to free a small brachiopod. During this parturition the fossil is exposed to freezing, thawing, heating, cooling, miniature landslides, and perhaps the misplaced foot of an animal. When it is finally free it is little more than dust. This is one case in which rapid weathering by man is better than the patience of nature.

A classic locale for silicified fossils is in the Glass Mountains of Texas, where a layer of Permian limestone contains ornate brachiopods in perfect preservation. Many of these brachiopods possess delicate spines projecting several inches from their small shells. These are invariably broken off when the fossils weather from rock layers. Blocks of this Glass Mountain limestone, taken back to the laboratory, placed in glass aquariums, and covered with acid, will in some weeks or months be reduced to a mound of magnificent fossils, perfectly cleaned and so delicate that they often collapse of their own weight when the water is drained away. A 180-pound block of this stone, when dissolved, was found to contain 10,000 brachiopods, an average of nearly 55 a pound, as well as other fossils. Large blocks are favored for this treatment, as they are more likely to contain large undamaged specimens. Such giant pieces are beyond the capacities of the amateur fossil collector, but pieces of a pound or so should produce fine results for him.

Another classic locale for silicified fossils is in the Ordovician of Virginia, where acid dissolution of blocks of limestone has released trilobites, particularly juvenile forms never before found.

An easy way to learn whether loose fossils found in an outcrop are silicified is to break off a corner of one and scrape this corner against a coin. If it digs into the metal easily and leaves a distinct gouge, it is silicified. If it doesn't make a good scratch, it is probably calcified. Often a weathered fossil, even if silicified, will wear a thin coat of limestone that will fizz and give a false signal in acid.

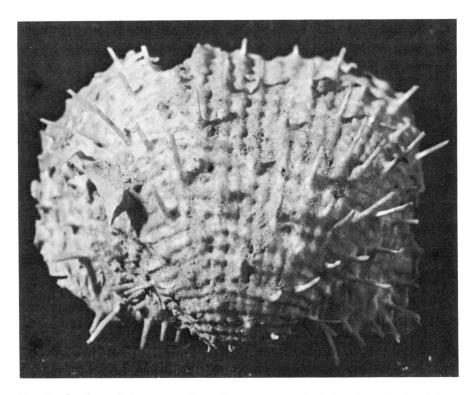

Permian brachiopods in western Texas limestone are silicified and can be freed from matrix with acid. Specimens so treated often show the original spines and delicate details that would not survive weathering. *Dictyoclostus;* Glass Mountains, Texas.

The block of fossil-bearing stone to be dissolved should be placed at the bottom of a glass or plastic bowl and covered with acid. If all the fossils seem to be on one side of the block, that side should be placed facedown, and only enough acid to cover the fossil layer should be added. The acid must be changed daily, as it quickly exhausts itself. The action goes on slowly after the violent bubbling stops. This violent bubbling can rip apart newly exposed, fragile fossils. If the fossils are particularly delicate, it may be wiser to use very weak acid for a longer time.

The mud that most limestones contain does not dissolve. It settles to the bottom of the container along with the fossils and makes their recovery from this murky mess difficult. If the specimen is placed on a piece of plastic screening supported above the bottom of the dish by glass marbles or quartz pebbles, the mud will settle through the screen, leaving the fossils behind.

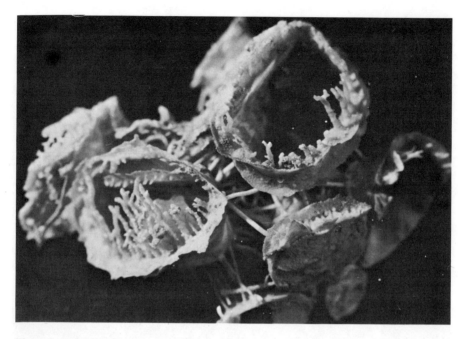

These strange fossils with the delicate spines are unusual brachiopods whose shapes closely resemble those of the horn corals. Such delicate fossils are collected by dissolving away the limestone matrix with acid. *Prorichthofenia uddeni;* Permian; Glass Mountains, Texas.

After the block is gone, the fossils should be washed gently in water. If they are in mud, it should be washed away carefully, preferably not down the sink, as mud may clog the drain. The fossils should be dried on a piece of paper toweling. If they are very tiny, they should be rinsed in acetone or alcohol to remove water from them. Since most water is hard because of dissolved minerals, the small fossils will glue themselves to paper or glass or whatever they dry on as the water evaporates, leaving behind a layer of its once-dissolved minerals. This water glue is surprisingly tenacious; the tiny fossils may break before they snap loose. Fossils should never be dried over an open flame, since water still in them may form steam and burst them apart. Tiny fossils heated in a pan will dry to a certain point and then suddenly pop and splatter out of the pan.

Pyritized fossils may be exposed with hydrochloric acid. Most pyritized fossils are found in shales, however, and unless there is a high content of calcium carbonate in the shale it will not be touched by acid. Devonian fossils often become pyritized; the brachiopods of Sylvania, Ohio, and western New York State are examples of this process.

Pyritized fossils are often found in rock layers near layers of coal. Many so-called pyritized fossils are really replaced by marcasite, the unstable sister of pyrite. To expose the fresh surface of a marcasite fossil by acid action is to invite disaster. In a few months or years the marcasite may grow white whiskers, produce sulfuric acid, and crumble into a pile of corrosive dust and acid. This has happened to many pyritized snails found in the coal mines at Farmington, Illinois. The cause of this marcasite disease has recently been attributed to the appetites of iron-loving bacteria. Treatment to thwart them consists of soaking the specimens in a strong bactericide for a day or more and allowing them to dry without rinsing.

Thick-shelled fossils often are not completely pyritized and if they are cleaned in acid, they can be damaged. The acid will enter the inside cavity of whole brachiopods and eat out the limestone, leaving pyrite shells a few millimeters thick that fall to pieces. Cleaning pyritized specimens with acid will always remove some surface detail. While producing a golden fossil, such cleaning materially reduces its scientific value.

All acids except hydrofluoric can be flushed down the drain after use if

Fossils replaced by marcasite or pyrite often develop a white-whisker disease from dampness. This is caused by bacteria and can be halted with a strong bactericide. *Shansiella carbonaria;* Pennsylvanian; Farmington, Illinois.

they are followed with plenty of water. Acids will not harm plumbing but will etch concrete basement washtubs. They will not damage porcelain.

Other acids, such as the dry powder acids, have limited use in fossil preparation. They are no better than the liquid acids and are usually expensive. If you are ordering acid from a catalog, you can choose the cheapest; it is not necessary to have chemically pure acid for dissolving rock. Some dilutions are cheaper than nearly pure acid, but sometimes the reverse is true. Remember that the two commonly used acids, acetic and hydrochloric, are easily available as vinegar (acetic) and muriatic (hydrochloric).

In using acid, remember:

1. *Always Add Acid* to water (AAA)
2. Always test a sample piece first.
3. Put a drop of acid on a spot of matrix before plunging the whole specimen into acid.

X

SPECIAL TECHNIQUES

Every collector needs to know something about techniques that are more complex than the methods of preparation we have described so far. These are now-and-then matters—the preparation of certain unusual fossils that require specialized treatment. As the collector becomes more skilled in his work with fossils he will wish to go beyond the scraping, chipping, and other methods that had formerly sufficed.

Such techniques as the making of peels, thin-section work, and plastic embedment are not beyond the ability of the advanced collector, and they will add new satisfactions to his hobby and better specimens to his cabinet. Turning a dull-brown coal ball into a series of montages of the cellular structures of ancient trees, leaves, and seeds only 1/1000 inch thick gives him not only personal satisfaction but also a closer look at the life of ages otherwise hidden by the mists of time.

THIN SECTIONS

Thin sections, used in paleontology as in biology primarily for identification of plants and animals, are extremely fine slices of material prepared by grinding or other means. They are sheer enough to allow light to pass through them, making cell walls, pores, and other details visible. They must be viewed through a microscope.

Another type of thin section, known as a serial section, is used to determine the shape of a fossil embedded in matrix from which it cannot be removed, such as a seed in a coal ball or a brachiopod in marble. The thin serial sections are removed one after another, progressing right through the fossil. The outer margins of the fossil, visible in the thin sections, are measured and plotted on paper until the three-dimensional shape of the fossil is reconstructed. This is time-consuming. The same result can be accomplished by grinding away a measured amount of the specimen in stages and drawing the fossil outline at each stage on paper.

Thin sections can themselves be things of interest and beauty. A peel of a coal ball may show in fascinating detail a cross section of the intricate cell structure of a cone or a root. Thin sections necessary for identification of petrified wood are also attractive for their colors. They can be mounted like pieces of color film, in glass projection slides and projected on a screen. Slides are particularly well suited to sections of petrified wood and coal balls, though interesting sections have also been made of coral.

Grinding Thin Sections

Thin sections can be made by two techniques—grinding and the peel. The first is like grinding a gemstone. The second uses acid to etch away a thin surface layer of matrix, leaving behind the actual cell walls of the plant or animal. These are then covered with a layer of liquid plastic. When it is dry, the plastic layer is pulled off, tearing loose the cells but keeping them in exactly the same position in the plastic.

Grinding of thin sections is commonly done with silicified wood, which, like modern unfossilized wood, requires a cross section and sometimes a longitudinal section for identification. This is true also of fossil corals and bryozoans, which are identified largely by the position, shape, and size of the inner chambers. Sometimes these animals can be identified without making a thin section, but the fossils must still be cut and polished in a flat section both across and lengthwise to expose the chambers.

Most books used for identification of fossil corals (particularly horn corals) and bryozoans show sections of the individual species, sometimes without a picture or drawing of the entire animal, since so many species are quite similar on the outside.

Before grinding a thin section (or preparing a flat area on a whole specimen) it is usually necessary to cut the specimen with a diamond saw. A thin blade should be used, particularly with tiny fossils. Normally, a section should be taken across the specimen (like cutting a carrot into slices) and another at right angles to it (like cutting the carrot lengthwise). Before sawing the specimen, study illustrations of sections in a textbook to see how to cut the specimen and what a section should look like when sawed.

Only a small piece of the fossil is needed, rarely more than half an inch. Don't try to saw it too thin; it may break.

Most fossils are too small or too irregular in shape to be held in the saw vise; they must be cut by hand. The saw should be drawn slowly and carefully to avoid breaking off the tip of the slice at the end of the cut. Since a diamond sawblade will not cut flesh, there is little danger from the saw unless a thin slice shatters while being held.

The sections should be washed at once to remove the coolant oil. Undiluted liquid detergent can be rubbed over the specimen. It can then be rinsed in water that is neither very hot nor very cold, as extremes of temperature can cause a specimen to crack.

If the saw cut is smooth, the piece can be cemented directly onto a glass microscope slide. If both surfaces are rough, one should be carefully ground or sanded flat. This can be done more easily if the surface exposed by the first cut is sanded to a flat surface before the thin section is sawed off. This leaves at least one flat surface on the slice.

Cement the thin section to the slide with epoxy or one of the fast-drying jewelry cements. Whatever cement is used, it must dry clear. Spread the cement evenly and thinly on both specimen and slide and put them together carefully to avoid trapping bubbles. Gentle heating at 200 degrees for fifteen minutes will cure epoxy sufficiently so that the specimen can be ground; without heating, it must cure at room temperature for a day.

The mounted piece should be ground as thin as possible. Ideally this should be done on a small flat lap, a cast-iron disc with a true flat surface. The disc rotates horizontally like a potter's wheel. The lap wheel is mounted atop an arbor, and the whole is placed in a metal tub to catch the grit cast off by the rotating lap. A motor, usually mounted underneath, turns the wheel at a speed determined by the diameter of the wheel. The speed should be just short of that which will sling off the grit as fast as it is placed on the lap wheel. These flat laps can be purchased for reasonable sums from a rock shop or built from an arbor and a lap wheel.

If a great deal of material must be ground away, the lap wheel should be sprinkled with 220 silicon carbide grit. If a large surface on a piece of petrified wood is to be prepared, 100 grit may be necessary for the first grind. The thin section should be ground until it is almost thin enough, which will be perhaps 1/25 inch. At this stage, the specimen should be translucent when held to a bright light. The lap wheel should then be cleaned, and the grinding continued with 400 or 600 grit. If the piece is

Cast-iron flat lap grinds flat sections for microscope examination or for preparing coal balls or petrified wood.

thin, and the fossil is composed of a soft material like calcite (most animal fossils), the grinding can be done entirely with 400 or 600. Grinding should proceed slowly, and the flat must be inspected often to avoid grinding through the fossil. The specimen will be complete when it becomes paper-thin and quite translucent.

An alternative method is to use a silicon-carbide grinding wheel, such as is used for lapidary work, and to grind gently on the flat side of the wheel. Specimens of more than an ounce or two should not be worked in this way. Some pieces of lapidary equipment, particularly the compact machines, are designed to run horizontally, making flat-section grinding easier.

Do not use much pressure while grinding on the side of the wheel, as it is not designed for side pressure. Use plenty of water for cooling and use a fine-grit wheel, such as 220. Move the specimen back and forth to avoid grinding a hollow in the wheel.

When using the grinding wheel, leave a slightly thicker section than when using the flat lap. The grinding wheels can also be used to remove enough material to proceed directly to the 400 or 600 grit on the flat lap, saving one lapping step there.

If no flat lap is available, the specimen can be given the final grind by hand on plate glass with loose silicon carbide grit. The grit (400) is sprinkled on the glass and enough water added to make a soupy mixture. The slide and its attached specimen are then ground vigorously with a circular motion. As soon as the specimen is ground to the proper thinness, it can be finished on another piece of glass with 600 grit. A piece of wet 400- or 600-grit silicon-carbide paper attached to a hard, flat surface such as a linoleum tile or sheet of glass can also be used for finishing.

It is not necessary to polish the thin section. A thin cover glass is usually cemented over it to provide a durable surface. The cementing can be done with a thin layer of epoxy. The cover glass should be pushed around until all bubbles and excess cement are squeezed out. Cover glasses can be bought at any store that handles laboratory equipment. They are less than a millimeter thick and must be handled carefully.

After the cement has hardened, the slide is ready to be examined under the microscope. A label should be attached, describing the section and the specimen it came from.

If an entire specimen is being ground in section, the operations are carried out as with the thin section. It is far easier to work with the whole fossil, as there is no danger of grinding through it. A cover glass can be glued with epoxy to an appropriate spot to make a window, or the specimen can be kept wet while being examined. This gives the appearance of a polished surface and makes the structures much easier to see. The surface should not be polished, because polishing pulls out soft areas and makes a rough surface rather than a smooth one. Pieces of solidly silicified wood can be polished, using standard lapidary procedures, after a good fine grind has been achieved.

After the thin section is made, or a suitably oriented flat spot is ground on a specimen, the internal arrangement of compartments can be compared to the illustrations in an identification book. It may be easier to make a pencil drawing of one area of the specimen rather than refer to the microscope repeatedly.

The Peel Technique

Peels are used mainly to examine coal balls, those mysterious mason jars of perfectly preserved vegetable matter 250 million years old. Coal balls have been described in Chapter IV—they are rounded masses of seeds, leaves, stems, roots, and bark of Coal Age trees "petrified" in complete cellular detail by a mass of calcium carbonate. These coal balls lie in the coal seam which was formed from similar plant material. The coal balls, impregnated by calcium carbonate, hardened into stone before the rest of the plant remains were compressed into coal.

Early peels were made by pouring liquid plastic over the flat acid-etched section of the coal ball. When the plastic was dry, usually hours or days later, it was pulled and scraped off the specimen. This process is rarely used now except when extremely small structures must be preserved.

The modern method is to use a sheet of cellulose acetate dissolved by acetone directly onto the specimen. It flows into all cellular spaces left after the matrix has been dissolved by acid. As the acetone evaporates, the plastic hardens again and can be "peeled" in less than an hour.

The peel technique can be used on some corals, brachiopods, and other fossils. The fossil must contain some acid-resistant organic substance. Most plant petrifications do contain such organic material, even the hardest silicified woods. The acid dissolves the matrix, which is calcite (calcium carbonate). Even agatized woods can be peeled if the silica is dissolved with hydrofluoric acid. A short bath in acid removes a layer of matrix perhaps 1/1000 inch thick, leaving the cell walls and other organic material standing free. The plastic fills the voids and tenaciously holds the organic material in exactly the same position when the plastic is ripped loose from the specimen. The peel needs no further preparation to be examined, identified, or displayed.

The specimen must first have a flat surface. This is usually made with a diamond saw. A coral may need to be oriented carefully in order to obtain a useful peel, but a coal ball may be cut into thick slabs at random. (It is impossible to know what structures will be where in a coal ball.) After sawing, the sections are ground perfectly flat, like a thin section, on a flat lap, or with loose grit on a piece of plate glass. The specimen is finished by grinding it on plate glass with 400 grit for a few minutes to make a smooth surface. The grit should be washed from the specimen.

A solution of 10 percent hydrochloric acid is prepared (Always Add Acid to water) in a shallow plastic, glass, or enameled flat-bottomed container. At least an inch of acid should cover the bottom. Grasp the specimen with rubber gloves and hold it, prepared surface down, in the acid, but do not let it touch the bottom. Any fine exposed structures will be broken off if the specimen touches the bottom. It should fizz violently. The time of the bath will vary from five to fifteen seconds, depending on the type of matrix, the strength of the acid, and how many times the acid has been used. It does wear out. Rinse the specimen by letting warm water flow across it gently for ten seconds or so. Be careful not to bump or touch the etched surface.

Place the wet specimen in a box of sand or gravel with the etched surface up and parallel to the floor. It can be allowed to dry for an hour or so or the drying can be reduced to a few minutes if acetone is poured gently across the surface a few times. This should be done in a ventilated place to avoid breathing the acetone fumes. While the specimen is drying, cut a sheet of cellulose acetate to a size somewhat larger than the specimen. The acetate is a clear plastic and should be about three mils (3/1000 inch) thick, or about the thickness of heavy paper. Do not confuse cellulose acetate with polyethylene or Mylar; it should be ordered specifically from a company found under Plastics in the Yellow Pages of the telephone book. If none is available locally, it can be ordered from the Colonial Kolonite Company, 2232 West Armitage Avenue, Chicago, Illinois 60647. Acetate comes in large sheets that can be cut to size.

Acetone should now be poured over the specimen. It evaporates rapidly, so it is necessary to work quickly or the peel will be incomplete. Start the acetate sheet at one corner of the specimen, the lowest corner if the acetone is running off that end. The sheet should be held slightly curved (see illustration) so that it is rolled across the specimen pushing a little wave of acetone ahead of it. Do not touch the specimen or wiggle the sheet to remove air bubbles. Allow it to dry for at least half an hour or until there is no more acetone odor.

When it is thoroughly dry, carefully pull on one corner of the sheet. It

Making a coal-ball peel. Step 1: After the coal ball has been sawed, it must be ground perfectly flat. A small amount of number 400 silicon carbide grit is used on a sheet of plate glass.

Step 2: The grit is moistened to a soupy consistency, and the coal ball is ground for several minutes with a rotating motion. It is then washed thoroughly.

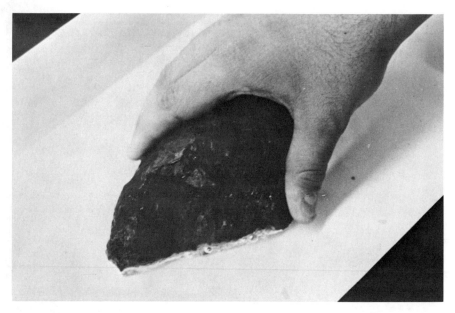

Step 3: The prepared surface is lowered into a shallow pan containing 10 percent hydrochloric acid. Immersion time varies from five to fifteen seconds. Care must be taken not to damage the prepared surface by striking it on the bottom of the pan.

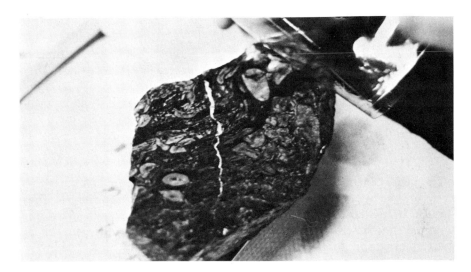

Step 4: After the acid bath, water is run gently across the surface to remove the acid. Care must be taken from here on not to touch the delicate, etched surface. The coal ball is propped with the top horizontal and allowed to dry. When dry, the surface is flooded with acetone, and a sheet of cellulose acetate is rolled onto it.

Step 5: The peel normally dries in half an hour. It can then be pulled carefully from the coal ball and is ready for study under a microscope.

should separate easily from the specimen. Pull it off slowly and gently, and the peel is complete. If another peel is desired, the specimen should again be ground for a minute or so on the glass with 400 grit to prepare a new flat surface, and the peel operation repeated. With care, more than a hundred peels can be made from $\frac{1}{4}$ inch of coal ball.

Peels made from corals, brachiopods, or bryozoans may have to be made several times to arrive at the optimum time of acid etch. Their small size may make the placing of the acetate film more difficult. It is advisable to etch silicified woods or coral with hydrofluoric acid, but only in a properly equipped laboratory.

The peels can be stored in envelopes, or particularly interesting structures can be cut out with scissors and mounted on microscope slides. They can also be placed in a projection slide, mounted between thin glass, and projected onto a screen. If some grains of matrix remain, the peel can be plunged into acid for a few seconds, washed carefully, and dried. The plastic is unaffected by acid. Names or catalog numbers can be written with a grease pen or marking pen on a corner of the peel.

Many coal-ball structures can be identified from paleobotany texts. Much specialized work has been done recently at the University of Illinois on such structures, especially of seeds. Workers there have published a number of papers illustrating coal-ball fossils. These are still available.

Peel sections of corals and other marine fossils are identified like standard thin sections. Most books dealing with identification of corals and bryozoans publish illustrations of thin sections.

PLASTIC EMBEDMENT

Liquid plastics are being used by hobbyists to create colorful wall decorations, jewelry, and tabletops with slabs of agate and other gemstones embedded in the glass-clear resin. This casting plastic is a polyester resin—a thick, sticky, slightly bluish liquid that hardens in a few hours at room temperature when several drops of a catalyst are added. The material can be poured into a plastic or ceramic mold; it can also be cut and polished after it hardens. It sells for $4 to $10 a gallon at hobby stores and rock shops.

Museums have been experimenting with similar substances, and the casting techniques they have developed can easily be adapted by the amateur.

Some fossils that are not dissolved by acid and that lie on the surface of limestone or limy shale slabs soluble in acid are excellent candidates for

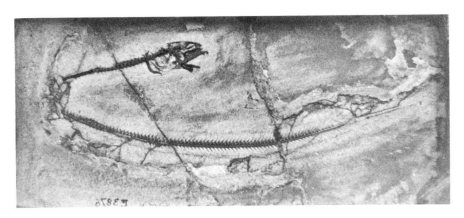

Fossil fish preserved in plastic after matrix was removed by acid. Specimen is in the British Museum.

making a plastic embedment. The method permits a view of both sides of very thin, filmlike fossils such as carbonized plants, carbonized worms, or graptolites. It is excellent for preparing fragile bony fossils, such as fossil fish. It can also be used for pyritized or silicified fossils that are so paper thin or fragile or so badly fractured that they would fall apart if freed. from the matrix. Shattered bones exposed at the surface of limestone can be kept intact with the plastic.

The specimen should first be prepared as well as possible on the exposed side while the other side is still locked in the matrix. All matrix clinging to the surface should be removed. The slab of rock should be sawed as close to the fossil as possible without risking damage to hidden parts or weakening the slab. If the matrix will not stand such treatment, or if saw oil and water would damage the fossil, the specimen should be trimmed by nibbling away at the edges with strong pliers or chipping it with a small hammer and chisel.

The surface of the trimmed block should be dry and clean. A piece of plate glass should be scrupulously cleaned and treated with a mold release (or wax such as Pledge) over an area larger than the fossil. The glass should be placed quite level, and a retaining wall should be built about an inch from the edge of the specimen on the surface of the glass. The easiest material for such a wall is an inch-high strip of Mylar plastic sheet. Only Mylar will work; polyethylene or cellulose acetate will not. It should be thick enough to stand by itself, about the thickness of heavy paper, which is about 5/1000 inch. Mylar is available at most stores that sell liquid plastics, or it can be obtained from large mail-order houses. The Mylar strip can be affixed to the glass with a leakproof band of modeling clay, self-adhesive rubber molding strips, or masking tape. The band is placed outside the strip.

Enough plastic should be mixed with the catalyst (following directions on the can) to cover the bottom of this modernistic pool to a depth of $\frac{1}{4}$ inch or less for small specimens, a bit deeper for specimens more than 6 inches in length. The layer should never be more than $\frac{1}{2}$ inch thick in one pouring—heat is generated as the plastic sets, and this may cause a thick pouring to crack. The plastic can be mixed conveniently in a disposable paper cup. The plastic should be stirred carefully to prevent bubbles from forming. If bubbles are trapped in the poured layer, they can be forced to the top with a toothpick.

Place a piece of paper loosely over the pool to keep dust from the newly poured surface. The material will become tacky and hard enough to support the specimen in half an hour to an hour, depending on the amount of catalyst and the ambient temperature. It is necessary to wait only until it has set firmly before adding another layer. Mix another batch of the plastic, and pour a thin layer—again less than $\frac{1}{4}$ inch—over the first pouring. Paint the fossil and matrix surface with a heavy coat of the liquid plastic, and carefully place the fossil facedown in the liquid in the mold. If one corner is submerged first and then the specimen is slowly rolled into the liquid, there is less chance of capturing bubbles. Lift the glass slab and peer underneath to see whether the piece is relatively free of bubbles; if there are many large ones, pull out the slab and try again. Add more plastic so that it rises around the sides of the specimen at least a half inch for small pieces, up to an inch for larger ones. Cover and allow to set for a day.

The top plastic may still feel sticky even after it has set for a long time. This is normal. The side against the glass will be hard and dry. Remove the side walls and try to separate the plastic block from the glass. If a release agent was used on the glass, it should come right off; if not, try a few gentle taps. It may be necessary to tap a table-knife blade under one corner to free the block. If it still sticks, place fossil and glass slab in the refrigerator freezer for half an hour. The plastic should then come off easily.

The fossil should be clearly visible through the thin window of plastic. Submerge the block in acid—10 percent hydrochloric for most fossils, and acetic for bones—in a plastic or glass container. The block should be placed with the fossil window facing up, and one side of the bottom blocked up on a piece of glass or other acidproof material. Refresh the acid when necessary, and allow the entire matrix block to dissolve. In the last stages it may be necessary to remove the block and gently wash the surface to remove clinging matrix.

Bony fossils may shed a few bones that were not attached to the skeleton and not exposed enough at the surface to be held by the plastic. When all matrix has been dissolved away, a skeleton should remain resting on the plastic, or a delicate carbon film of a plant or graptolite firmly fixed on the plastic. Rinse the block gently for half an hour in water to which a spoonful of sodium bicarbonate has been added. This will neutralize acid remaining on the fossil. Rinse the specimen again with clear water and allow it to dry.

When the fossil is *thoroughly* dried (bones can hold moisture for a day or more), more plastic can be mixed with catalyst and gently poured into the well and over the newly exposed back of the fossil. No more than $\frac{1}{4}$ inch should be poured at one time, and the plastic must be allowed to set between pourings. Few fossils will need more than one or two pourings. The plastic should be kept as thin as possible, just enough to cover the back of the fossil.

The final pouring can be covered with a liquid available from the store that supplied the plastic. The liquid will keep air from the surface and allow it to dry hard. If the plastic is still tacky, the piece can be heated gently in an oven or an electric fry pan until the plastic sets. You will notice that the plastic will set more rapidly on a warm day than a cold one. However, plastic should not be poured on an excessively warm and humid day, because the moisture will cloud the plastic.

The fossil is now preserved with all delicate details of both sides clearly exposed. Protruding edges of the plastic can be ground away on a wheel and sanded with sandpaper but the article should be allowed to dry for several days before it is handled or polished. Final polishing can be done with a cotton wheel and buffing compound, or on leather or felt sheets with standard polishing agents such as cerium oxide, tin oxide, or tripoli. If the surface becomes scratched, it can be repolished.

Casting resins do not have a long shelf-life. Make sure to buy fresh resin and keep it in a cool place, such as in a refrigerator. Once it is opened, use the contents of the container in a short time. Never mix more than can be used in fifteen minutes, because in that time it will set. Spic and Span or other detergents can be used to remove the inevitable sticky messes from the hands. Acetone is a solvent for the liquid resin but is not recommended for use on the hands. Work should not be done in the kitchen because the odor of the casting resin can affect the taste of food.

Liquid casting plastics have other uses in the fossil field. A specimen, such as a clam shell or a delicate snail embedded halfway in matrix but too badly fractured to be loosened further, can be coated with a thick layer of plastic and then prepared from the other side. Hand preparation can be used on shales and sandstones that would not be affected by acids. The

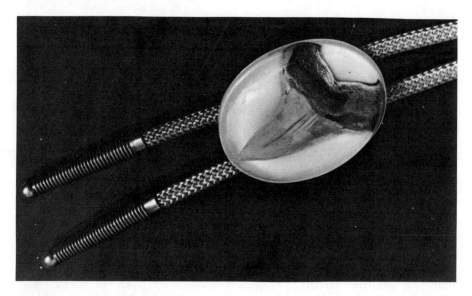

This fossil tooth once belonged to a shark; it now adorns a collector. A standard ceramic mold was used. Two pourings of casting resin were made: the first was of clear resin (done upside-down) with the tooth embedded, followed by a backing of white resin. With bola-tie slide attached, the fossil becomes a piece of jewelry. (Jewelry by Cecelia Duluk)

plastic acts as a glue and a solid base for the specimen. Extremly thin fossils, or fossils on thin sheets of shaly matrix, can be strengthened on the back with layers of the plastic. The plastic can also be used to cement a fragile but irregular piece of matrix to a wooden mount. Of course, fully prepared fossils can be embedded in the plastic, either in a polished block for a decoration or in a mold to make a shark-tooth bola tie slide, or a pyritized brachiopod pin. Commercial molds are available in ring, bola tie slide, and pin shapes.

CASTING FOSSILS

Most fossil collectors have been struck, perhaps at Christmas time, with the beauty of a fossil used as jewelry. Since most fossils will not wear well if cemented to a ring or pin backing, the obvious answer is to cast the specimen in harder material.

Casts of fossils are used in schools, where there is danger of damage to irreplaceable specimens that will be handled a great deal. Casts can be made by the hobbyist for donation to schools and interested junior collectors.

Some models made commercially are difficult to tell from the real fossil. One collector found a superb trilobite in a dusty drawer in a London mineral shop a few years ago and bought it for a reasonable price. Later, when he was washing the specimen, one stark white corner appeared on the otherwise brown trilobite. The specimen turned out to be a painted plaster cast, made perhaps fifty years ago.

The black, shiny, fat trilobites found in the Devonian shales at Sylvania, Ohio, are particularly well adapted to casting. A black plastic cast is startlingly lifelike. One collector cast a number of these trilobites and then on a club trip to the quarry scattered them in likely places before anyone else got there. All day there were shrieks of excitement as one person after another unearthed these perfect trilobites. Many found it hard to believe that the joke was on them. At the quarry some weeks later a woman excitedly showed friends her prize find of the day. It was one of the plastic trilobites, speckled with mud from intervening rains. Only its light weight betrayed it. She was heartbroken when told the truth. Undoubtedly some of these fake arthropods, labeled in good faith as the real thing, reside in collections.

Casts of type-specimens of fossils are professionally made to be sent to other museums and universities for their study collections. A good cast is

Cast in plastic of the crinoid *Cactocrinus arnoldi,* made in a rubber mold.

in every way as useful as the real thing. It shows all details, even the smallest pit and pore. It does not show color variations and suggestions of organic films as well as the original does and, of course, cannot be further prepared or examined under a microscope to see crystalline detail.

Casts are also made in natural molds, which are the only fossils found in some rock layers. This is particularly true of dolomites. Sometimes a calcified fossil is found in an extremely hard sandstone, slate, or shale that cannot be chipped away without damaging the softer fossil. In such cases, the fossil is dissolved with acid, and the mold cavity filled with rubber. This yields a perfect cast of the fossil when the hard matrix is broken away.

A century ago, casts were complicated things to make. One method was to coat the specimen with shellac to seal the pores and then press it into wax or paraffin. By another method the fossil was lightly coated with oil and covered with plaster. The mold was removed from the fossil, and plaster casts were made from it. Often only one could be made. Fossils that had projections or overhangs had to be cast from elaborate split molds, sometimes half-a-dozen pieces for each specimen. Such molds were not only difficult to make but difficult to cast.

Relatively smooth fossils with no undercuts can be cast using plaster molds. The fossil is coated with oil and half buried in plaster (left mold). When thoroughly dry, the plaster mold is treated with a release agent such as petroleum jelly, and the rest of the fossil is covered with plaster. When dry, the two halves of the mold will separate easily from the fossil and from each other. A filling hole must be cut into the mold at the highest point of the cavity. The mold must be treated with a release agent before each casting. This blastoid was cast in plaster in this mold.

Casts made well over a century ago by James Hall turned up when his collection was unwrapped from ancient newspapers at the Field Museum of Natural History in Chicago, which acquired the collection. The casts were made of sulfur and of lead.

The amateur can make simple casts of fossils such as flat trilobites, small ammonites, flat snails, brachiopods, and other fossils that do not have exceptionally intricate detail or overhangs. This is easily done by coating them lightly with oil (which is not well adapted to absorbent or porous matrices) and pouring melted wax over them. Papier-mâché works well to cast fossils that are without fine detail, such as bones or shark teeth. Modeling clay will hold an impression of any fossil sturdy enough to be pushed into it. These molds can be cast in plaster but are usually good for only one cast.

Multiple castings from all except very fragile fossils can be made with rubber molds, the contribution of this century to casting. Hobby- and art-supply shop sell liquid latex in bottles. A pint bottle makes many molds. The fossil is painted with several layers of this material, and then the

John Harris of the Field Museum shows use of rubber molds in casting large fossils. He is touching the lower jaw; the plaster cast, a faithful reproduction, is in front of it. This jaw was cast with only two molds (at right side of picture). Such large rubber molds are good for making only about a dozen casts.

rubber mold is pulled from the fossil. Since rubber stretches, it can be used on fossils with some overhangs, or even on rolled trilobites or other three-dimensional fossils. If it can be peeled off the fossil without stretching out of shape or splitting the rubber, it can be used as a mold.

The rubber mold is filled with casting plaster, which is allowed to dry and is then removed. Rubber molds can be used dozens of times before they split. Even split molds can be repaired with liquid latex. Casting plastic (polyester resin) can be cast in these rubber molds, but it must be removed carefully, as the surface will be sticky no matter how long it cures. It can be hardened by gentle heating in an oven or electric fry pan.

The liquid plastic can be tinted with special dyes mixed in along with the catalyst. The surface of plastic casts may be shiny and may need to be dulled by a rubbing with fine grit. Plaster casts can be painted with watercolors to resemble the natural color of the fossil.

Rubber is the best material for making casts from natural fossil molds in the rocks. The rubber cast can be painted, or it can be used to make

Mold of a snail in dolomite. A release powder is dusted into the mold and casting rubber is poured into it. The rubber cast is much easier to study than the fossil mold itself.

another mold that can be cast in plaster or plastic. Unless the collector leaves some sort of handle protruding from the rubber when he pours it into a natural fossil mold, he will wind up with a hole full of rubber and no way to pull out the cast.

Molds have been made of silicone rubber and of polyvinyl chloride, a type of plastic. But the amateur will not have use for them. Molds designed for making lead or other hot-metal casts require special techniques.

If a simple natural mold is seen in a rock, a field cast can be made by coating the mold lightly with thin oil, pressing modeling clay into the cavity, and gently pulling it out. Even though distorted, the mold will pick up fine details and disclose what once occupied the hole.

X-RAY EXAMINATION

The use of X rays is a recent development that has become the standard method of identifying mineral specimens. The use of X rays in identifying fossils is less well known. It is beyond the capacities of most amateurs except persons such as physicians who may have access to such machines; but the technique should be understood by all collectors so that they can avoid destroying a rare fossil by attempting to prepare it by other methods, when it would have been a perfect candidate for an X-ray picture.

X rays reveal fossils locked deep within hard matrix just as they show bones in a human body. Under the right conditions, a fine X-ray picture can be made of a fossil that is not visible at the surface. Preparation might have been impossible because of the nature of the matrix or so difficult that there would have been much damage.

X rays produce a picture because there is a difference in the absorption of the rays by bone and flesh or fossil and matrix as the rays pass through the object to the film. Absorption of rays increases to the fourth power of the atomic number of the atoms that compose the fossil and matrix. Thus, if a pyritized fossil is embedded in a carbonaceous shale, a fine X-ray picture can be made. Pyrite is an iron sulfide, and iron has an atomic number of 26. Carbon has an atomic number of 6. This difference allows strong contrast because of the much greater absorption of the X rays by the pyrite.

On the other hand, a pyritized fossil inside an ironstone concretion will make a very poor X ray. Both fossil and matrix contain iron. Limestones are hard to X-ray unless very thin because they absorb all X rays. A calcite fossil would not show up; it is chemically the same as limestone. Carbonaceous shales with pyritized fossils are excellent subjects. This is fortunate, because they are the most difficult fossils to prepare mechanically.

Dr. Eugene Richardson and Dr. Rainer Zangerl of the Field Museum in Chicago did a massive research project on a fossil occurrence in central Indiana—a mass burial ground for Pennsylvanian fish. The fossils required slow, painstaking preparation. There were thousands of specimens. After much experimentation and the addition of an electronic dodging machine to compensate for the wild contrasts in the negatives, superb X-ray pictures of the fish were produced. The pictures were more useful than even the best-prepared specimens. This technique is described in *Handbook of Paleontological Techniques* (see Appendix: Recommended Books.)

If an amateur finds black, sheety shales having suspicious swellings and some exposed scales or bones in them, which he is unable to prepare, he may find they are of great interest to a well-equipped museum.

There will probably be other developments in this area. Research is being done with neutron radiography, which can be used to produce pictures similar to X rays. A neutron radiograph of an object containing a variety of substances, such as wood, plastic, and several different metals,

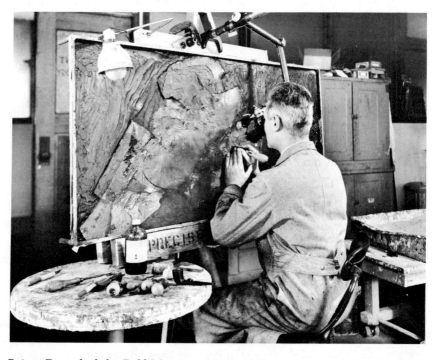

Dr. Rainer Zangerl of the Field Museum of Natural History preparing fossil shark in hard shale. Broken fossil was reassembled in a frame and prepared with small scrapers and flaking tools. Pennsylvanian; Mecca, Indiana. (Photo Field Museum of Natural History)

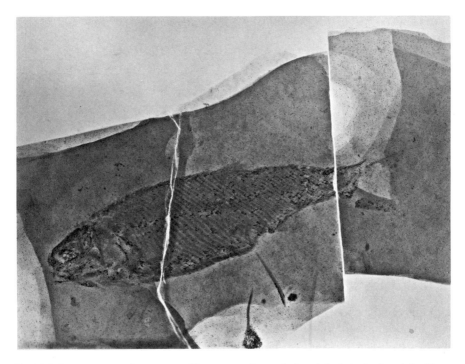

Fossil fish apparent only as a swelling on surface of the slab but disclosed in X-ray photograph in perfect detail. Pennsylvanian; Mecca, Indiana. (Photo by Eugene Richardson)

would show each hidden object distinctly in a different shade of gray. A standard X ray would not differentiate as well and would ignore some of the objects.

While this has not yet been applied experimentally to fossils, it undoubtedly would give much better results than X rays with fossils and matrices. Its drawbacks are the high cost of the equipment, which is now limited to nuclear laboratories, and the danger to users and passersby from radiation.

FLUORESCENCE

While examining a collection of the Jurassic fossils from Solnhofen, Germany, a researcher was startled to see a beautifully detailed insect fluorescing brilliantly on one slab. When he turned on the room lights, the insect could not be seen on the piece of limestone. Other disappearing insects were found on other slabs. The conclusion was that the insect-shapes were indeed fossils, but fossils whose substance had decomposed before the rock hardened. No swellings or visible fossils remained, but organic fluorescent substances once contained in the insects were left behind, creating ghostly images of their bodies. Visible fossils were found to show details not visible before—for instance, shrimp showed antennae and legs under the ultraviolet light that could not be seen under the most careful scrutiny in daylight.

This was an exceptional case of fossil fluorescence—the insects that weren't there. Some fossils fluoresce, but none as brilliantly as minerals do; and the minerals that replace the organic material of fossilized organisms are not fluorescent at all or only feebly so. Consequently, the cause of the fluorescence of some fossils is a challenge to science.

It has been suggested that, as with the Solnhofen fossils, the fluorescing material is a substance remaining from the organism. A few snail shells display fluorescent bands and stripes, perhaps the remains of long-vanished color patterns.

Just as crabmeat canners separate the fluorescent crab shell from the nonfluorescing meat by ultraviolet light, so the collector occasionally may use the same technique to distinguish small fluorescent fossils from the matrix. In the same way, he may find that fossils from one locality closely resemble those from another, but those from one locality are fluorescent and the others are not. Here he has a ready means of separating them.

One further use is akin to the use of ultraviolet light to detect alterations and forgeries in art works, rare stamps, etc. Most glues and cements used in patching fossils are fluorescent, so that the lamp may be a means of detecting even the cleverest repair work.

The ultraviolet lamps, both shortwave and long, that are used for minerals are satisfactory for fossils. They come in a wide range of prices and sizes, varying degrees of portability and powers.

XI

CATALOGING
AND DISPLAYING FOSSILS

A jumble of fossils on a shelf is nothing but a heap of curiosities of neither scientific nor monetary value. Among them may lie a one-of-a-kind fossil, a new species, perhaps even a "missing link" that could throw light on the relationships of some group of plants or animals. Or a common fossil may take on new interest because it is from a locality where such a species has not been found before or from a rock series in which it has never been seen previously. But if the fossil has not been carefully labeled and cataloged with the exact location and rock layer in which it was found, it has lost all significance. Recording, labeling, and displaying are the steps by which the collector educates himself as well as contributes his bit to the science of paleontology.

LABELING

All fossils should be labeled as soon as possible. Fossils taken from separate formations should be separated in the field. Fossils from one location should not be mixed with those from another, even though the two localities are only half a mile apart. The fossils may be from similar but distinct formations. Furthermore, discards from one site should not be dumped at another site; they should be saved for the driveway. Dumping them thoughtlessly may thoroughly confuse research on that locality.

Most fossils found by the amateur will be loose—weathered free from the matrix. If they are collected in matrix, the blocks of rock containing the fossils will probably be found some distance from their source. This makes it difficult to say with certainty which layer contained the fossils. They should be so labeled.

Few amateurs have to go through the sweat and tears of mining out layers of rock to be sure that fossils are found in place. This is what the professional must do; the loose fossils so readily available are only colorful objects to him. If a fossil is taken from a specific layer of rock, the general description of the rock layer should be noted on the catalog card for the fossil; for example, a record might read, "thin, three-inch gray limestone lying below a one-inch black shale and three feet above a thin coal seam."

The location information should be as specific as possible. This could be the name of the quarry, a specific spot in a river (such as "100 yards downstream from crossing of State 41"), roadcut (described by milepost or mileage from important junction or river crossing), or beach cliff (distance from some prominent feature). The best description is by precise position within a township. This would have to be done by reference to a quadrangle map (see Chapter VIII). An exact description will outlast road changes, river meandering, fluctuating city limits, and filled quarries. There are many fossils in old museum collections with descriptions of locations such as "300 rods southwest of Mills Ferry Crossing." A town now thrives where Mills ferried a century ago, and nobody now knows where the crossing was.

A label can be written on a piece of paper and wrapped with the fossils in the field, or the bag or box can be labeled. As much information as possible should be included at the time; memory fades fast.

CATALOGING

As soon as the fossils have been prepared at home, they should be cataloged. Each specimen should have a catalog number, even though twenty from the same locality are kept for display or research. All can be given the same number if they are identical, or each can be given the same number followed by a different letter.

There are as many systems of numbering as there are collectors. The easiest system is to start with the first fossil as No. 1 and number specimens as they are received. Another is to start with a letter followed by a number: the letter referring to the fossil phylum, period, or locality, and the number to its order of acquisition.

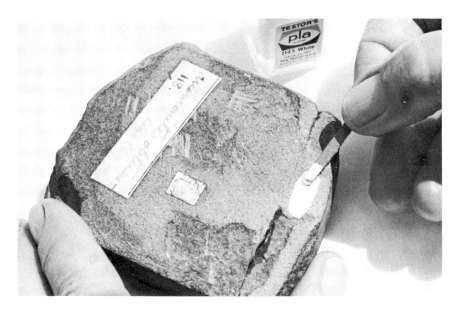

Fossils are numbered by painting a white spot on the fossil with lacquer or enamel.

After the paint has dried, number the fossil with India ink. When it is dry, seal it with a dab of household cement. Save old labels, such as this 19th-century English one.

The most permanent method of labeling a specimen is to write the number on the specimen in some obscure place. On light-colored matrix this can be done with a pen fitted with a fine nib (such as a crow-quill pen) using India ink. The number can be written directly on the matrix. The India ink will be permanent. On dark matrix, such as black shale, white ink can be used, but this is not as permanent.

For uniformity, a splotch of white can be painted on the specimen to receive the number in India ink. The white enamel or lacquer sold in most hobby shops for model painting is excellent. A strip of paint the width of a paper match and a quarter inch long will easily hold a four-digit number; in fact, a match makes a good throw-away paintbrush.

The paint is dry enough to write on in ten minutes. If a mistake is made, the India ink can be washed off immediately. After the ink is thoroughly dry, at least fifteen minutes, paint a thin layer of plastic household cement (such as Duco) over the number. Do this rapidly in one stroke, as the cement will smear the ink if it is brushed on. The cement protects the number from wear and from repeated washings.

On exceptionally rough matrix, such as sandstone, a thick layer of household cement will smooth the spot before painting.

Paint the number on the matrix rather than directly on the fossil. If the actual fossil must be numbered, do it in a spot that will be hidden when the specimen is displayed and in an area not vital for proper identification. Tiny fossils may not have room for a splotch of white. If they are not porous they can be numbered directly with ink.

The number should be entered in a catalog along with pertinent data. Some collectors prefer a double system of bookkeeping, with brief description in one book for rapid referral, and another book or series of index cards containing full data. An entry should include names of the fossil, precise location where found, age and formation when known, date collected, and any comments of interest about the fossil. It is helpful to include a description of the fossil, listing size, condition, and even approximate value. If a specimen is purchased, the dealer's name and price should be included. Having two separate catalogs will be added protection if one is lost. Whenever making an entry, try to enter a description that would make sense to the museum curator who may acquire the collection many years later.

DISPLAYING

Entire books could be written about techniques of display. Museums hire artists and decorators to prepare exhibits that are educational and interesting to the public. The amateur may keep his collection tucked away in drawers or shoeboxes, or he may display it so as to amaze and delight his neighbors and instruct the local Cub Scouts.

Cabinets and Drawers

Most fossils are small—several inches long or less—and are unimpressive if hidden behind other specimens on wide shelves. The usual wooden shelves, china cabinets, or department-store display cases are not well designed for small fossils, which are best seen in an open tray, such as shallow drawer or a flat case not more than a few inches deep covered with glass. If an upright case must be used, it should not be deep, or the fossils in the rear rows will be poorly lighted. Since few fossils have the overpowering color and beauty of fine minerals, most visitors to a basement museum prefer to look at only a few cases of particularly interesting specimens. The rare ones of interest only to collectors can be kept in drawers.

Museums and universities house collections in cabinets that may stand thirty drawers high. Each drawer is about twenty-four inches square and two inches deep. Few specimens except some cephalopods, giant corals, and vertebrate fossils are too large to fit into such a drawer. Drawers can be built at home or can be bought in sections from geological-supply houses (see Appendix). Dentists' and typesetters' cabinets and map-storage units for libraries occasionally appear on the market. They are excellent, although a bit shallow. Drawers have many advantages. They afford a maximum of usable display space; they take up little space themselves and they keep specimens free of dust. Furthermore, specimens can be examined closely in drawers and even removed for closer inspection, if desired.

Specimens on large thin slabs can be hung from nails or pegboards by hooks attached to their backs with tape or epoxy.

Flat display cases can be built from shallow drawers or from two-by-one-inch lumber (furring strips) attached to a plywood backing. A thin railing of wooden molding strip can be attached inside the case an eighth of an inch below the top, just deep enough for a sheet of thin glass to be flush with the surface when resting on this railing. A small hole in one side allows a finger to slide under and lift up the glass when it is to be removed for cleaning or for changing specimens. These cases can be built in a variety of shapes and sizes to sit atop tables or benches.

Small fossils can be displayed in the cardboard cases known as Riker mounts. These are a foot or less long and an inch deep. They are filled with cotton, have a glass cover, and are not very expensive.

Museum-type method of arranging and storing fossil collection in shallow trays placed in drawers. Each tray contains card with place, date, and collector of the fossil, and vertical card with catalog number, name, and other data for quick reference. Collector is shown comparing a specimen ammonite with reference book. (Photo California Division of Mines and Geology)

Backgrounds and Supports

Fossils are not colorful, and a good display will take advantage of background material and props to please the eye. Flat cases and drawers should be lined with some material that will help stop round fossils from rolling every time the drawer is opened. The best background for most fossils is a dark material of fine texture, unless the fossils are very large. Black con-

Styrofoam blocks make attractive mounts for dark-colored fossils. The front of the block can be beveled and a label attached with pins.

struction paper is cheap and can be cut to fit a drawer without overlapping. Black cloth is even better.

In china cabinets or other large cases with shelves, specimens will have to be supported in position. This can be done by cutting the matrix into a block so that the specimen sits upright, or the fossil can be pressed into a piece of styrofoam. Small blobs of non-oily clay or Play-Do can be molded to the bottom of the fossil as a base. Oily clays will transfer their oil to the specimen; it will soak irretrievably into the fossil. Sometimes a small piece of styrofoam or wadded paper put behind the specimen will hold it at a proper angle. There is no need to glue a specimen to a block of wood, plastic, or styrofoam unless it is for a permanent museum display.

Grouping Fossils for Display

Fossil displays are usually arranged either by groups of plants or animals or by age. Trilobites would be in one section, fossil ferns in another, and brachiopods in a third. Fine displays have also been created by grouping fossils from one locality. In competitive displays at rock-club shows and geological-society conventions, fossils are often exhibited in other classifications, such as fossils personally collected; fossils of one genus or family; fossils showing replacement by other minerals; fossils of a certain size (such as microfossils); or fossils displayed to show variety of forms within one small group. These themes make home displays more interesting than endless rows of battered brachiopods collected over the last decade from the local gravel pit.

Repairs and Retouching

For display purposes, some fossils may need to be touched up to enhance their good features. But they should only be touched up, not repainted, refinished, or remodeled. Even after careful preparation, a fossil may have so little color contrast with the matrix that it is hard to see. This is particularly true of fossil leaves and soft-bodied animals that are only a film on the rock. Such specimens should never be coated with shellac or paint to bring out contrast, as this may destroy the fine features needed for identification.

Many plant fossils from Illinois and Indiana have been daubed with thick, shiny varnish or shellac, which not only decreases contrast but also leaves an annoyingly shiny surface that is impossible to remove.

Fossils that need increased contrast should be coated with yellow dextrin, an inexpensive substance formerly used in baby food. It can be obtained at large drugstores or at some rock shops. White dextrim will not work; demand the yellow. A pinch of the yellow powder dissolved in a teaspoon of hot water will coat several dozen fossils. Apply the dextrin with an artists' paintbrush, and be careful not to paint the matrix. The coating should darken the fossil without leaving a particularly shiny surface. If it is too shiny, the amount of dextrin in the mixture should be decreased. Mistakes can be washed away in warm water.

Some specimens benefit from an all-over bath of the dextrin, while others look better with the matrix painted and the fossil left untouched. Experiment, but if you can't do it with dextrin, you can't do it with paint, varnish, or plastic, either.

Now and then, specimens that have been unfairly restored appear in collections. Black carbonized-fern fossils missing some of their detail have been restored with touches of India ink. Normally dark-brown or black trilobites have been made more so with a lustrous coating of shoe polish. Missing spines of some fossils are scratched into the matrix and colored. Plates of unusual fossil combinations are concocted by gluing specimens in desired places. Whether such things belong in a fossil collection is a question that must be left to the conscience of the collector.

Some small repairs are permissible. It is fair to cover up glue joints in broken specimens. When repairing a broken fossil, take care not to allow

Before treatment, this fossil fern lacked contrast with its matrix.

The fossil, painted with weak yellow dextrin solution, shows improved contrast. Matrix is not painted.

surplus cement to show at the surface. Dust the exposed top of the crack with scrapings from the fossil matrix; when the glue dries the crack will be camouflaged. Holes and chipped areas can be refilled with a dough made of epoxy and rock dust of the proper color. Fossil bones missing some portions or skeletons missing some bones often are filled into their original form in museum preparations. Plaster is used to do this, and usually the museum is careful to color this plaster so that it is noticeably different from the real bones.

PHOTOGRAPHING FOSSILS

Many fossil collectors will never wish to take pictures of their prizes, but some will want to make color slides to illustrate lectures or black-and-white prints to submit with articles. They will find that photographing fossils is a specialized aspect of the craft.

Most fossils are only an inch or two long, too small to be photographed with inexpensive fixed-lens cameras that cannot be adapted for closeup lenses. A good 35 mm. single-lens reflex camera with a fast lens and a set of extension tubes or a bellows is necessary for color slides. It also works well for black-and-white print photography. Larger cameras, from $2\frac{1}{4}''$ x $2\frac{1}{4}''$ to 4″ x 5″ press cameras, are excellent if prints are wanted.

Portrait or closeup lenses are sold for attachment to cameras with fixed lenses. These are not expensive and will produce good pictures of large fossil specimens. Some lenses bring the camera into focus within a few inches of the specimen, but unless the photographer is a good guesser, the fossil will rarely be centered in the film.

The single-lens reflex camera has certain advantages for this type of work. It allows sighting directly through the lens that will take the picture instead of through the auxiliary viewing lens which is an inch above the real lens. In distance shots, both lenses see essentially the same picture, but in closeups a small fossil that appears directly centered through the auxiliary viewing lens may not even appear in the final picture. The single-lens reflex shows things exactly as they will appear on the film, even including the depth of field.

Numerous books that explain how to take closeup pictures of small objects are available in camera shops. One of these books should be consulted for its tables about magnification achieved with combinations of different portrait lenses and extensions of the lens from the camera body. As the camera lens is moved farther away from the film, the image size becomes larger and larger until a small fossil completely fills a negative. By the use of bellows or extension tubes two- or threefold magnification can be achieved, which is more than ample for normal photography. Higher magnifications can be made by using a microscope with camera attachment.

Two things happen when extension tubes or bellows are used. The exposure time increases dramatically but according to a definite formula that will be listed in any book about closeup photography. The depth of field also decreases, until at the magnification needed to take a picture of a one-inch brachiopod with a 35 mm. camera only a fraction of an inch will be in focus. This means that much of any rounded fossil will be out of focus. Small lens-apertures of f/32 or f/64 help increase this slim depth of field, as do special lenses designed for closeup photography. A book or a professional photographer should be consulted about these.

Fast film and a sturdy tripod are needed for color-slide photography. If a lens extension or portrait lens is used, the exposure time, even with fast film and bright sunlight, will still be too long to get a sharp picture with a hand-held camera. Ektachrome is the favored film, but Anscochrome works well. Kodachrome is excellent, but slower. European color films work well.

Outdoor Lighting

Whenever possible, photography should be done outdoors to take advantage of the light, which is at its best for fossils in the morning and late afternoon. Around noon, the overhead sun does not cast shadows in the low spots of a specimen, and without this modeling the fossil will look flat and drab. Very early or very late sunlight should also be avoided for color photography because it creates too many shadows and is excessively orange.

The camera should be mounted atop the tripod at a convenient level. A table or platform is needed to bring the specimen up to the camera level. Some specimens should be photographed on a background of textured cloth, while others look better "floating" above an out-of-focus background. This effect is created by placing the fossil on a large glass shelf supported at least two feet above the ground. The ground is covered with cloth or paper of an appropriate color. It will be out of focus in the picture. The glass shelf, unless dusty, will be invisible in the picture.

Reflections are a problem when using glass, and the photographer may wind up with a picture of a fossil sitting inside a reflection of the photographer taking the picture. After sighting through the lens to make sure no reflections are seen in the picture, the photographer may move his head a few inches to snap the picture and inadvertently create a bad reflection.

Few fossils will pose well by themselves. They need to be coaxed into position by props made of modeling clay or wadded paper. The surface of the fossil should be placed so that it is parallel with the lens to make sure that all of it will be in focus.

Since camera movement is magnified as much as the fossil in closeups, long exposures require use of a cable release. The movement of the mirror during exposure in a single-lens reflex camera creates vibrations, and this is compounded if the shutter is released manually. A breeze can jiggle camera or specimen enough to make a blurry picture; so can a passing truck.

If the camera has a behind-the-lens meter, the exact exposure time and lens aperture can be accurately and immediately read. If the meter is located elsewhere on the camera or is separate, allowance must be made by increasing the exposure to compensate for use of bellows or extension tubes. This can be determined by experiment or from tables supplied by camera manufacturers.

Pictures can be taken indoors using artificial lighting, but for the occasional photographer the bother of acquiring the necessary equipment and of setting it up is hardly worthwhile. There will be many days when good pictures can be made outdoors. Brilliant sunshine is not needed; a hazy day or one with thin clouds is fine, as such lighting softens the contrast between the shadows and highlights and seems to make colors richer. Dark shadows can be reduced on a sunny day by holding a piece of frosted glass, cloudy plastic, or thin fiberglass matting above the specimen to diffuse the sunlight. Dark shadows can be filled in by placing a piece of aluminum foil or white paper to the side of the specimen so that sunlight is reflected into the dark areas. Exposures should be figured with these devices in place.

Black-and-white photography is the most highly developed form for fossils. Prints made for reproduction in journals, magazines, and books must be of high quality because the print will lose some detail even with the best reproduction.

Improving Contrast

Fossils rarely have an even color. When photographed, blotches and stains look like shadows and obscure the real shape of the fossil. A dark specimen is difficult to light: there must be sufficient contrast between the highlights and shadows to show the three-dimensionality of the specimen. A white fossil photographs best of all, so a blotchy fossil can be whitened temporarily for photography. There are three different methods of doing this.

The easiest way, and one that is used professionally, is to coat the specimen with magnesium oxide. This does not harm the specimen, because it can be washed off after the fossil is photographed. A piece of magnesium ribbon several inches long is held by one end with a long pair of tweezers or pliers. The free end is lit by holding it in a gas flame or cigarette lighter for a few seconds. Magnesium, a metal, burns with a brilliant, intense flame and gives off clouds of white smoke. This white smoke is magnesium oxide.

The specimen should be held by one corner with a pair of tweezers and placed several inches above the burning magnesium strip. The strip burns for only a few seconds, so the work must be done quickly. It is wise to wear sunglasses to cut down the brilliant glare from the burning magnesium. If the specimen is held in one place, essentially parallel with the smoke column rather than at right angles to it, more of the white oxide will be deposited on one side of the bumps and corrugations of the fossil surface than on the other side. This delineates these features and in the photograph will look quite natural. The coating is so fine that no surface detail is lost, and if the result is not satisfactory the specimen is easily cleaned and resmoked.

A more controlled deposit of fine white powder can be made with ammonium chloride. A bottle of dilute hydrochloric acid and a bottle of ammonia (ammonium hydroxide) are connected with glass tubing that runs through a rubber stopper in the top of each bottle. Air is forced into the first bottle by blowing through a separate tube, driving the combined vapors of the two chemicals through another glass tube (see illustration). The combination creates ammonium chloride, a fine, white powder that is deposited on the fossil held in front of the exit tube. It is easier to control this chemical fogging than that from the rapidly burning mag-

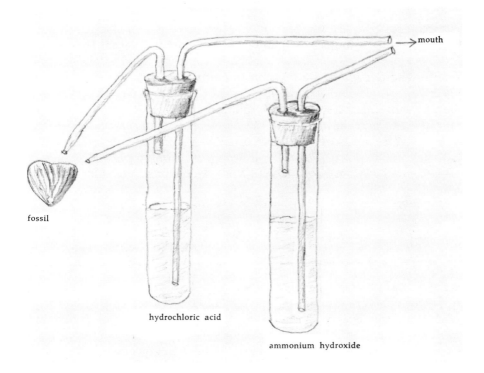

mouth

fossil

hydrochloric acid

ammonium hydroxide

Fossil is whitened for photographing by blowing fumes of hydrochloric acid and ammonia simultaneously through both tubes. The white powder forms as the fumes meet at the fossil.

nesium ribbon, though the apparatus is clumsy. Excessive humidity will cause the ammonium chloride coating to disappear. After the photograph has been taken, the fossil can be washed clean.

A third method, the least satisfactory but better than none, is to dip the specimen in thinned white watercolor paint or nonpermanent white ink. This may sink into the specimen and ruin it; try a scrap fossil before whitewashing a good one.

The ultimate in contrast is produced by dipping the specimen in India ink before coating it with one of the whitening agents. The whitening chemical is directed onto the specimen so that only the highlights are whitened, leaving the low spots coal-black. A fossil treated this way will clearly show every tiny pore and ridge in impressive contrast. Unfortunately, the India ink remains a permanent part of the fossil.

Photographing Fluorescence

Fluorescence of fossils can be photographed. Inexpensive black light bulbs are adequate light sources for photographing most fluorescent fossils, such as the mysterious insects from Solnhofen, Germany, or the color patterns in snails and clams that are invisible in normal lighting. There is an added advantage; in ultraviolet light the depth of field is markedly increased, allowing more of the surface of a rounded specimen to be in focus. Fast panchromatic film must be used for black and white pictures, but even then exposures are extremely long, often over thirty minutes. Filters such as Kodak Wratten 2A, 2B, or K2, Ilford Q, or Corning Noviol C must be used. Color slides can be made with Kodachrome or similar films, without using filters.

Infrared Photography

At the other end of the invisible wave lengths of light is another interesting field of photography—the infrared. Contrast is often much improved in fossils photographed in infrared light. Fish scales or graptolites are transparent in infrared light, and crustaceans in dark shales photograph much more clearly. This contrast cannot be seen by the naked eye, so the fossil must be photographed in order to find out whether contrast is indeed improved.

Any tungsten bulb (regular light bulb) is a satisfactory source of infrared radiation, but a camera filter such as Kodak Wratten 25A or 87, Agfa 85, or Ilford 207 must be used. Special infrared film is available, and the film must be developed in the dark. No safelight may be used. Cameras must be in perfect condition, as the bellows or wood used in some cameras may leak infrared light even though they are impervious to visible light. Since infrared light focuses at a different spot from visible light, the lens should either be moved slightly farther away from the film after focusing in normal light, or a very small lens opening should be used to compensate for the change in focus.

Developing and printing photographs of fossils is carried on just as with other pictures.

Photographs for Publication

For publication, most editors prefer a glossy print with good contrast that is at least five by seven inches in size. Several fossils can be photographed together on one plate if they can be kept in focus. Lighting should always be from the upper-left corner (this is a standard accepted by all professional paleontologists). When submitting photographs to a magazine, be sure to include in the mailing envelope a thick piece of cardboard as large as the largest photograph. This will dissuade the postman from folding the package when he stuffs it into the mailbox. Write clearly on the envelope: *Photos—Do Not Bend or Fold.*

Never write captions or identification on the reverse of the prints; the writing will show through as raised script when the print is reproduced. Identification information should be on a separate sheet but not stapled or clipped to the photograph. Paper clips press into the photographic paper under the weight of other letters in the post office. If writing must be done, do it on the margin of the photograph, preferably with a crayon or a grease pencil.

Be sure to note the degree of magnification or reduction of the fossils as they appear in the photograph or to include the actual size of the specimens in the notes. If a one-inch fossil is three inches long in the photograph, this is expressed as x3; if a three-inch fossil is only one inch long in the photograph, it is x$\frac{1}{3}$. It is helpful to consult the *Journal of Paleontology* to see how a typical article is illustrated.

TRADING AND SWAPPING

When the collector gets home, unpacks his trophies, and cleans and labels them, he will often find that he has duplicates of some species. He is now in a position to augment his collection by trading with other collectors.

Trading can be done at the many mineral- and fossil-club shows held throughout the nation, and at the conventions held annually by the six major mineral and geological society federations. Swaps can be arranged between persons who have met at such shows or who have obtained each other's names by mail or by advertising in the several amateur hobby magazines (see Appendix).

Material to be swapped is always more appreciated by collectors who live near the locality where it has been found. They understand its desirability and from personal experience can estimate the time and effort that has gone into collecting it. But exceptionally good material is likely to find a market anywhere.

The rule is to bring to rock swaps only good, clean fossils, marked with as much information as is known about each specimen, including its scientific name, age, formation, the locality where it was collected and the date when it was found. Material neatly arranged in flat trays or boxes with paper collars around the specimens to keep them unbruised will attract the swapper's eye. Some swappers mount specimens on styrofoam blocks using a silicone rubber cement which holds them firmly but can easily be peeled off by the new owner. This allows the specimens to be packed tightly in a box without requiring wrapping. It is well to take several grades of specimens, however; some to be moved quickly to keep the swapping active without too much regard to value, or to give to junior collectors, and some to hold onto until highly desirable specimens can be had in exchange.

A canny swapper looks over other swappers' material with an eye for what he wants, then brings out his best material. He must learn, however, to say No when some charming child or seemingly pathetic oldster tries to cozen him out of a prize specimen, or when something that he does not need or want is pressed on him.

An amateur fossil collector, mindful that he collects what he needs for personal use, swaps with other amateurs, but he buys from dealers and does not attempt to profit from his material.

XII

MICROFOSSILS

Collecting and preparing microfossils brings out the engraver in some collectors, for they must learn to handle, mount, and identify fossil organisms smaller than grains of salt. The world of the ultra-small has been extended to the tiniest fossils of all, the coccoliths, which become visible only with scanning by the electron microscope. These little shell-like fossils are so small that many species are invisible under a good optical microscope and require a magnification of 40,000 power to make them clearly visible in detail. A collection of coccoliths could be mounted on the head of a pin with plenty of room remaining for a label, but nobody has yet solved the problem of how to mount them.

A microfossil is any fossil that requires magnification to be identified properly or to be seen clearly. For all practical purposes, this means a fossil less than $\frac{1}{8}$ inch in the longest dimension. Few people collect these organic miniatures, probably because of lack of information about them, difficulty of identification, and the impossibility of convincing anyone else that a speck of dust is as interesting as a 6-inch trilobite.

Professional paleontologists have become greatly interested in certain mini-fossils in the past decade, as paleontology swings from the adventurous science of digging up dinosaurs to the laboratory drudgery of identifying conodonts and other microfossils.

Despite their size, microfossils have uses, particularly to the oil geologists, for they tell the driller when he has gone past where the oil should have been. In another hole, microfossils contained in the slender cores of stone removed from the hollow drill tell the geologist where he can expect a productive oil or coal horizon. Big fossils, the brachiopods and trilobites, are too rare to be conveniently centered in a drill core where the geologist happens to break it. So he must rely on fossils that are widespread, extremely common, and lived only during short periods of geologic time. These are the three qualities of an index fossil, which, when found, date and locate that rock layer. Thus, in the interest of higher profits to the mining and oil industries, micropaleontology has been highly developed.

Representatives of all major groups of the larger macrofossils are also found as microfossils. Many juvenile forms of larger invertebrates, such as crinoids and trilobites, are microscopic. Fascinating collections of these juvenile forms can be made and matched with their grown-up counterparts. Thousands of species of diminutive pelecypods and snails lived on the ocean floor. Brachiopods contain a few small members, as do the cephalopods and corals. Only the dinosaurs and large mammals have no real microfossils, except for thin sections of their bones, but the smaller mammals and fish have micro-parts, such as teeth, bones, and scales. There are several large groups of animals that have left no fossils except microfossils, notably the conodonts, ostracods, and foraminifera.

CONODONTS

Conodont fossils resemble a strange jawbone set with modernistic teeth. The largest run a dozen to the inch; most are half that size, still visible enough to the naked eye so that they can be picked up on the tip of a needle and mounted on a slide or can be seen on the surface of a slab or loose in the field. They are composed of solid calcium phosphate, making them a bit heavier than the quartz, dolomite, or calcite sediments in which they are found.

For quite a while conodonts were in a never-never land of classification, believed to be tooth structures of fish, worms, cephalopods, crustaceans, or snails, depending on whom you talked to. Because conodonts were often found along with vertebrate marine fossils, such as those of fish, many paleontologists tried to place them in some vertebrate order. But the elusive conodonts never came attached to any animal or even part of an animal, although they exist by the millions in almost every Paleozoic sedimentary rock.

At the 1969 meeting of North American paleontologists at the Field Museum in Chicago, the real conodont appears to have stepped forward to reveal its identity. In a surprise program presented by William G. Melton of the University of Montana, the fossil of a strange wormlike creature two to four inches long with conodont structures inside it was shown to the assembled scientists. The conodont animal, if it is such, was described as a soft-bodied, bottom-dwelling organism found with other coal-age organisms, mostly fish, in the Little Snowy Mountains of Montana. Whether it is the conodont animal, with a food-grinding mechanism in its midparts, or just a worm that swallowed a conodont assembly, has provided the paleontologists with a challenging new problem that may solve the old problem of the conodont's identity.

Conodont fossils are found in rocks ranging from Ordovician to Triassic, primarily in shales rich in organic matter, but are also found in some sandstones and limestones, particularly those interbedded with thin layers of shale. Some gray shales contain as many as 500 conodonts a pound, but the average for shale or limestone is more like 10 to 50 a pound.

Scolecodonts somewhat resemble conodonts but are composed of chitinous or horny material, similar to a fingernail. These are denfinitely worm jaws, found on occasion in association with fossil worms. They are not nearly as common and widespread as the conodonts, but some Paleozoic rocks do contain a fair number of them.

Left to right: Leperditia sp., a crustacean about 1/8 inch, Ordovician through Pennsylvanian; scolecodont, worm jaw about 1/2 mm, ranging from Cambrian to Recent; conodont, toothlike microfossil about 1/2 mm, ranging from Ordovician to Triassic. (This and the following drawings by Betty Crawford)

OSTRACODS

Ostracods are tiny fossils that look exactly like an ornamented clam, complete with both valves. The largest members of this widespread and rather common group of fossils border on macrofossil size, a few being the size of a small bean, though most are in the size range of the conodonts, 1/10 inch or less in length. Unlike the conodonts, which disappeared sometime in the Triassic, the ostracods first left their fossil record in the Ordovician and are still swimming about today.

Ostracods are members of the class Crustacea, subclass Ostracods. This makes them small brothers and sisters to lobsters, shrimp, and trilobites and no relation at all to clams or brachiopods, which they closely resemble. If one thinks of the ostracod shell as being like the chitinous covering of the crab, lobster, or shrimp, though divided strongly down the middle to form a top and bottom, with most of the organs tucked inside, it is easier to recognize ostracods as relatives of the other crustaceans. They are not bottom crawlers like crabs or lobsters but swim in both fresh and salt water.

Actually, the sea is full of tiny crustaceans, but few of these leave a fossil record because their protective shell disintegrates easily. It is made of organic material rather than the stony calcium carbonate of the brachiopods and clams. Ostracods deposit a layer of calcium carbonate beneath their more typical crustacean outer shell, and it is this part that becomes the typical ostracod fossil. The fossil shells are clamlike in shape, ornamented with bumps, ridges, or projections, generally light in color, and are found in vast numbers in some rock layers.

Any fossil-bearing layer of limestone or shale, but particularly limestone, is likely to contain numbers of these little creatures. Since they dissolve in acid, they cannot, like the conodonts, be removed from a calcareous matrix. Sometimes, by sheer luck, they will be found exposed on top of a limestone slab or weathered free in the soil. Shales containing ostracods can be disintegrated to release the specimens. The fossil record contains a vast number of species of these little things, which are widely used in rock dating, as are the conodonts.

FORAMINIFERA

Foraminifera are also found in the washings of shales, or more commonly, lying loose by the thousands near weathering limestones and chalks. They are generally in the category of microfossils, though a few very large specimens are the size of a coin. One even reaches a diameter of four inches, but most are about the size of a grain of rice or less. Forams, as they are usually called, are members of the phylum Protozoa and are found in prodigious numbers in present-day seas. Modern descendants live only in salt water, though some are found in brackish water that contains only a trace of salt. A few adventurous types live far underground in the slightly saline groundwater of wells in central Asia and northern Africa. It appears that all forams found as fossils were ocean dwellers.

The first fossil forams are found in the Ordovician, but they are not particularly common until they burst forth with tremendous prodigality in the Pennsylvanian and Permian periods. They are not rare as fossils in any rocks since the late Paleozoic but are again found in explosive numbers in Cretaceous rocks, sometimes making up the bulk of thick layers of limestone. In general, the post-Paleozoic forams are larger than their early counterparts.

Forams built outer skeletons (called "tests") of several hard materials. These are ideally suited to becoming fossilized. Most of these little animals built tests of calcium carbonate, but a few, particularly the early ones, built a test of carefully selected sand grains, sponge spines, or even tests of smaller forams. Each species was fussy about its building blocks, selecting only one type of sand (quartz grains, calcite grains, or even mica grains) or one particular shape of sponge spicule or foram skeleton.

The forams that constructed their houses of such carefully selected building blocks look like a rather baggy lump, usually rounded but some-

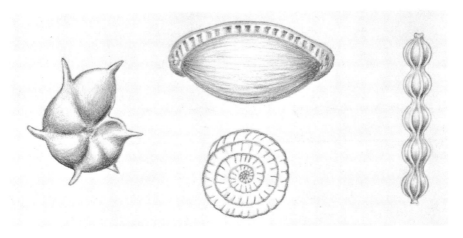

Three foraminifera. *Left to right*: *Hantkenina*, 1/2 mm, ranging from Eocene to Miocene; typical fusulinid, side view and cross section, about the size of a wheat grain, ranging from Mississippian to Permian; *Nodosaria*, 2 mm, ranging from Triassic to Recent.

times irregular or elongated. The calcareous ones are far more interesting to the micro-collector because they have definite shape. The typical calcareous foram looks like a miniature ammonite, especially in cross section, with a spiral arrangement of chambers. Some look like a rattlesnake rattle or a rose. They cannot be mistaken for any other microfossil.

Fusilinids

One special group of forams—the fusilinids—were abundant during the Pennsylvanian and Permian and then became extinct. The fusilinids, members of the family *Fusilinidae*, look like grains of rice on the outside and have a series of tiny chambers in a spiral on the inside. Most are about the size of a grain of rice. Chert filled with these tiny fossils is known as "rice agate" in Iowa and is used as a gemstone. Some species are larger in size, up to half an inch in length, but retain the rice-grain shape. These fusilinids are common in some rock layers lying in a belt from Iowa to Texas, where they weather out by the millions. They are particularly at home in limestones but are not common in shales.

RADIOLARIANS

The radiolarians are not widespread as fossils in the United States, but in some areas they are abundant and particularly show microfossils. Like the forams, they are members of the phylum Protozoa, and the order Radiolaria. Of all microfossils, they have the widest range in time, from the pre-Cambrian to recent times, though in this country they are found commonly only in Devonian cherts of Texas and California and in Jurassic rocks of California. The radiolarian skeleton is made of silica and, like the snowflake, often has threefold or sixfold symmetry, with projections also much like a snowflake. Others looks like a symmetrical wishbone. A few look like a cross or the head of a medieval mace. These creatures can be found in silica rocks, particularly quartzites and sandstones.

DIATOMS

Rather similar to the radiolarians in shape, but far more intricate, are the dazzling beauty queens of the microfossil world, the diatoms. They, too, have a skeleton of silica, but of opal rather than quartz. Opal is an unstable form of silica, and their skeletons were often damaged or destroyed as the opal changed to chert. Where they are well preserved, as in thick beds in California, the diatomaceous earth is mined for use as filters in such industries as brewing.

The diatoms make excellent filters because their incredibly intricate lacy skeletons are full of microscopic holes. Typically rounded if from marine deposits and elongated if from fresh water, the tiny skeletons have in places piled up into a solid layer of rock many feet thick. No fossil diatoms have been found in rocks older than Upper Cretaceous, but because they were quite intricate and highly evolved by this time it is suspected that their origin goes back much further in time.

Diatoms are plants, not animals: strange plants that still live in both fresh and salt water. The lacy opal skeleton is the support of the tiny plant, which is visible only under high magnification. The fossils are symmetrical and have thousands of tiny openings, always in matched pairs, piercing the clear opaline matrix. They seem to try to outdo each other in complexity, and many are somewhat reminiscent of snowflakes.

Because they are so tiny, diatoms are not likely to be picked up in the field except where a layer of diatomaceous earth is being mined. It is a simple matter to take a chunk of this white substance, crumble it, and examine the dust for desirable specimens. A piece the size of a fist may contain not hundreds, or thousands, but millions of diatoms. They are too small for an amateur collector to mount singly, but a little sifting of the dust on a slide should provide hours of fun at the microscope.

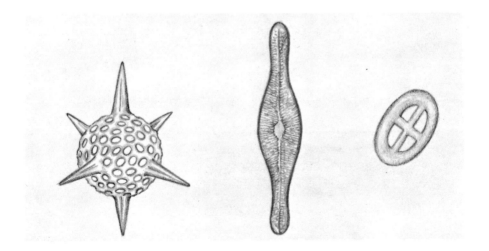

Left to right: a radiolarian, ranging from Cambrian to Recent; a diatom, ranging from Jurassic to Recent; a coccolith. These microfossils are all greatly magnified.

OTHER MICROFOSSILS

Other fossils are found solely as microfossils. Some are valuable tools for the professional paleontologist, telling him not only the age of the rock but what the climate, or sea temperature, or oxygen level of the water was at that time. Fossil spores and pollen are now widely studied, but this field is of little interest to the amateur collector.

Although every major group of invertebrates and plants has some species that can be considered microfossils, there are strange fossil occurrences of normally large-sized creatures found instead as small, but apparently mature, miniature versions of themselves. A Lilliputian community of this kind is called a depauperate fauna. The reason for the communal disregard for normal size is not at all clear. Such depauperate areas are not large but are widely scattered in time and space. Famous depauperate zones are found in the Ordovician of Iowa and the Mississippian of Indiana. In both places microfossils of cephalopods, snails, and several other types can be collected, many only a quarter normal size.

COLLECTING AND PREPARING MICROFOSSILS

Except for these depauperate zones in otherwise normal fossil-bearing localities, there is no particular place to look specifically for microfossils. They can be found in most rocks that carry bigger fossils, since the small

Collecting microfossils is best done horizontally. Dr. Dwayne Stone demonstrates the proper position. A small plastic bucket is an ideal container for this kind of collecting. (Photo by Betty Crawford)

creatures lived among the larger ones. There are some rock layers that have only microfossils, particularly those of such organisms as diatoms or protozoa. In fact, there are few marine sedimentary rocks that do not contain some microfossils.

Collecting these little fellows can be done by hands-and-knees scrutiny of weathered shales and limestones in the field, popping the tiny fossils into a pill bottle. This will do for some larger microfossils; but to make a real collection, chunks of unweathered rock must be disintegrated to release their tiny captives. This may be done by crushing the rocks, washing the residue, and inspecting it for unbroken individuals. Some shales fall apart easily when alternately wet and dried or chemically treated. A few species of microfossils not composed of calcium carbonate can be released by dissolving blocks of limestone with acid. The resulting sludge is washed, dried, and spread thin on a microscope slide for sorting.

Shales of dark color are likely to contain conodonts and scolecodonts, particularly if there is much organic material in the shale. Gray shales are often the home of small invertebrates or juvenile forms of larger animals. Many sheety black shales cannot be disintegrated, but most shales will crumble into their original silt and clay if properly treated.

Cleaning

Boiling with the Quaternary-O, as described in Chapter IX under cleaning techniques, will release fossils of conodonts and other mini-creatures as well. If this chemical is not available, satisfactory results can be obtained by gently boiling the shales in water in which trisodium phosphate or sodium hydroxide has been added. The shales will turn into mud, releasing the microfossils.

The disaggregated shale is bulky compared to its fossils. Before pouring off the water used to boil the shale, examine its surface carefully for floating fossils. Some forams are hollow and bob about like corks. They could easily wind up in the sewer instead of the cabinet. Then the chemicals can be poured off and the muddy residue flushed several times with fresh water. The water should be allowed to remain still for at least a few seconds before it is poured off, in order to allow the fossils to settle to the bottom. If this washing is done carefully, much of the clay still in suspension can be poured away with the water without losing too many fossils. Pouring this clay-laden water down the sink may stop up the drains or sewer. It is better to pour the muddy water into a bucket and then on the ground.

When it seems inadvisable to pour off any more water, the residue should be washed with distilled water. Most water contains some dissolved minerals which would be deposited around the tiny clay grains and micro-fossils when the water evaporates. For a tiny fossil this mineral water acts as an effective glue, cementing it to fellow fossils, mud grains, or the side of the container. Fragile fossils so fastened will break in pieces before coming loose. But distilled water removes most of the chemical glue. For faster drying, the residue can then be soaked with alcohol or acetone after the washing with distilled water.

The final drying can be done on a sheet of metal such as a cookie sheet, or on a piece of filter paper or smooth white paper. Do not force-dry the mud. If it is heated it will not only splatter, but tiny fossils may explode when the water included with them turns to steam.

When the sediment has dried thoroughly, it can be spread thin, a small amount at a time, on a microscope slide and examined under low power for fossils. These can then be removed for mounting.

Limestones can be dissolved to release conodonts and scolecodonts, both of which are impervious to gentle acids such as 10 to 15 percent acetic acid. Pyritized or silicified microfossils obtained by dissolving the host rock are much better specimens than those obtained from the surface of a natural exposure. The pieces of limestone are placed in plastic or glass bowls and covered with the acid. The acid is refreshed when it appears

Disintegrated shale spread on white paper is examined for microfossils under a low-power microscope. Pointer touches a Pennsylvanian brachiopod from Danville, Illinois.

that its action has stopped after some hours or days, depending on the quantity of stone and acid. Be careful not to move the stone when pouring in new acid; tiny fossils protruding from the matrix at that point can easily be broken. Do not cover the bowl tightly. Carbon dioxide gas is produced during the destruction of the limestone and can build up dangerous pressure.

The sediment remaining after acid treatment of limestone must be carefully washed with many changes of water. There is usually not much excess sediment, such as there is with shales; so the water should settle for at least fifteen seconds between changes to allow all fossils to sink to the bottom. A pinch of sodium bicarbonate tossed in the water of one of the last changes and allowed to remain for a few minutes will neutralize any acid remaining in the fossils. From this point, the residue is treated like the shale residues, washed in acetone or alcohol, and spread out to dry.

Sorting Microfossils

Commercial sorting techniques have developed among those who spend most of their lives looking for microfossils in drill cores. But they are beyond the means of the amateur collector. Conodonts are composed of calcium phosphate, which makes them a bit heavier than clay or limestone particles, and they can be separated by using heavy liquids to float off the lighter sediment, which is useless. Conodonts are also weakly magnetic and have been successfully sorted out by jiggling the residues down a slight slope; a strong magnet on one side pulls the conodonts toward it and feeds them into a separate chute.

When thoroughly dry, the dusty remains of the amateur's washday are spread out thinly on a glass slide and examined under a low-power microscope or a powerful hand lens. When an outstanding or unusual specimen is found, it is transferred to a special microfossil mounting slide. This kind of slide is designed to hold a piece of black, ruled paper (white for dark-colored specimens) in a little well, with room above for a cover slide that will keep out dust and prevent stray fingers from dislodging the fossils. One such slide can hold several dozen specimens. A homemade mounting slide can also be used, but unless the specimens are protected they are likely to pop off and get lost. On ruled-paper mounting slides, a code number can be written next to each specimen for identification. It is also possible to mount several specimens atop a cork in a standard micromount box.

The microfossils are most easily removed from the debris on the sorting slide with the moistened tip of an artists' paintbrush. They can then be transferred to their permanent home and placed on a tiny spot of glue to hold them down. Several water-soluble glues are used professionally, but the amateur collector can cement the tiny things with the glue from gummed paper tape such as is used for sealing boxes. If an artists' brush is moistened, passed along the glue side of the tape, and then touched to the mounting spot of the fossil, the mucilage remaining on the mount will be sufficient to hold the fossil. This method works well for small, light microfossils. Heavier cements such as Duco or Elmer's are too bulky to work well. Professional paleontologists often attach the specimens with a solution of gum tragacanth.

Identifying Microfossils

Microfossils are identified just like their bigger relatives. There will be much difficulty in identifying juvenile forms of crinoids and trilobites, because these creatures are not commonly illustrated in texts available to the amateur. Professional journals and papers dealing with the locale where the fossils were collected may be of help. The easiest way to specialize in juvenile forms and identify them is to find a paper written about the juvenile fossils of a specific locale, then go to that locale and collect. Such articles appear occasionally in the *Journal of Paleontology*.

Young individuals rarely have more than a slight resemblance to their parents, but depauperate forms have the form and shape of their normal counterparts on a smaller scale. They can be identified with standard texts, as can those fossils that are normally found only as microfossils. Lately a wealth of material has been written about conodonts, ostracods, and other commercially useful microfossils. State Geological Survey bulletins and such magazines as the *Journal of Paleontology* are good sources of locales and identification for these tiny creatures.

Microscopes

As with micromount collections of minerals, the best viewing device for microfossils is the binocular microscope, which allows a three-dimensional view. A high-powered scope is not necessary; most microfossils need only about ten to sixty times magnification. Inexpensive (under $15) binocular scopes imported from Japan will do a satisfactory job, and a secondhand binocular dissecting microscope, such as is used in school biology classes, can sometimes be bought for $50. When purchased new, large binocular scopes (the Japanese ones are small) are rarely available for less than $100, and most cost more than $200.

A monocular scope is usually cheaper and gives satisfactory results. New imported small scopes cost $10, and a satisfactory used one of a larger type can be found for $25.

Most microfossils are not transparent. They have to be lighted from above rather than from underneath. Most microscopes have either a light under the stage that holds the slide or a mirror underneath that reflects light from an outside source. Since such devices are useless for most microfossils, a special light that gives an intense small light from above will be needed. Such a light can be purchased, but it is cheaper to buy a small high-intensity lamp that uses six or twelve volt bulbs. These little lamps usually are jointed so that they can be twisted to shine the light just where it is needed. The light should not be left burning too long above a specimen, or the heat it creates may pop the fossil from its mounting.

Storing

Prepared slides with microfossils can be stored in standard slide files obtainable at any store that sells microscopes. Since most slide files are designed to hold a slide vertically and since microfossil slides must be stored horizontally, the collector should either buy a horizontal holder or

Microfossils that are not too small can be mounted in a micromount box. This ray tooth is mounted on top of a blackened tooth pick. Identification is written on a label on the lid keyed by number to the box.

store a vertical one on edge. The tiny spot of glue holding a microfossil may break loose when the slide is vertical, and the fossil will be left hanging out in space. These slides should always be handled with utmost care.

Slightly larger microfossils can be mounted individually in a standard micromount box such as the kind used for minerals. These boxes are obtainable from most mineral supply dealers; some are made of black plastic, others of clear plastic that must be painted flat black to cut down light reflection. The fossil is mounted on a small cork or atop a toothpick or thin balsa rod, also painted black. A plastic cement such as Duco or silicone rubber can be used to cement the specimen to the mounting rod. The name of the specimen and locale are written on a label attached to the plastic cover of the box, and the boxes are stored in shallow drawers.

XIII

FOSSILS
AND THEIR FAMILIES

A collector's interest lies in his fossils, in the fun he has had obtaining them, and in the skills he has developed in preparing and displaying them. But fossils are something more than stones: they once lived, and the true collector will wish to understand generally where each of his fossils belongs in the grand scheme of life. When he does understand, his view is enlarged, and his fossils attain in his eyes the dignity and meaning to which their age and importance entitle them.

Furthermore, acquaintance with the major families of animals and plants and their peculiar patterns as fossils equips the collector to recognize basic habits and structures. For example, fivefold symmetry in a fossil would instantly alert the collector to the strong probability that it was one of the echinoderms—which include the starfish, sand dollars, and crinoids. A spiral or conelike shell with a single aperture would tell him that he had a gastropod, or snail; two like shells with a central hinge, that he had a pelecypod, or clam; unequal shells (in profile) with one hanging over the other at the hinge would mean a brachiopod, which is a very different breed of shellfish.

Such acquaintance with gross characteristics of the various phyla and classes is also necessary to enable the collector to make intelligent use of reference books. Instead of paging through a book until he chances on

Page from a report by James Hall, New York paleontologist, with sketch of fossil and woodblock from which it was printed. Inset is the fossil itself.

fossils resembling the one he is trying to identify, he will turn at once to the chapter dealing with the group to which his fossil probably belongs. He can understand the technical terms and descriptions he will encounter in such books.

This brief review of fossils by phyla in the order of their increasing complexity of structure is designed to help the collector make the best use of his time with his fossils and to deepen his enjoyment of them. Those who want to know more about the subject can consult the works listed in the Appendix: Recommended Books.

PLANTS

Plants had to come before animals, for plants in a sense create their own food, whereas animals cannot. The energy of sunlight is used by plants to form sugars, starches, cellulose, and other complex organic materials from carbon dioxide and water. This process, photosynthesis, is the very basis of life. It creates the foodstuffs that animals must have to satisfy their appetites, to grow, and to reproduce. Some animals eat other animals, but somewhere at the start of the line was a grazing animal that lived on plants.

The first cell must have been more primitive than any existing today, even the bacteria, but it did have life and the ability to divide and create another individual, and it must have found some way to create its own food as all plants do.

Well-defined remains of algae and bacteria nearly two billion years old have been identified in the Gunflint cherts of Lake Superior. Similar fossils are known from Canada, Africa, and other parts of the United States. They are not showy, and are of little interest to the amateur collector except as indications of the chronological vastness of his hobby.

But in the iron mines of Minnesota, especially at the Mary Ellen mine near Biwabik, he can collect a red jasper handsomely marked with swirl patterns. This is called algae agate, and the swirls are believed to be fossil colonies of a primitive iron-secreting organism.

Classification of Plants

The classification of plants in the scheme of things, like the nature of these earliest organisms, is not precisely agreed on. Some authorities are content to divide all organisms into the animal and the plant kingdoms. Others divide them into as many as five kingdoms: 1. bacteria and blue-

Lepidodendron tree bark, found in shales or sandstone overlying coal seams of Pennsylvanian age. Pella, Iowa.

green algae; 2. protists, which are other algae, protozoans, and slime molds; 3. fungi; 4. true plants; and 5. animals.

Paleontologists have difficult classification problems with plants. Parts of plants such as leaves, roots, and bark are usually found separately, and since their relationship is not apparent, receive names which later must be changed when discovery of a more complete plant throws light on the situation. *Lepidodendron*, for example was the name given to the trunk of the lycopsids; roots of the same genus were called *Stigmaria*; leaves were assigned to *Lepidophyllum*, and cones to *Lepidostrobus*. Yet all belonged together, just like a tooth, a bone, and a scale from a fish. There are still many similar puzzles; there are still many divisions of opinion about them among paleobotanists, and many names that may be changed.

Thallophytes

It is a simple solution to group all organisms that are not animals into one kingdom of several divisions. The first of these, the thallophytes, includes plants that have no circulatory system and no woody tissue. For this reason they must live in water or damp places. This division includes the microscopic bacteria—single-celled thallophytes which not only cause disease but also are nature's principal agents for breaking down waste organic matter. They are nature's scavengers.

Algae

Next above them in complexity are the algae. Great reefs of *Cryptozoon*, cabbage-like colonies of algae, grew near Saratoga Springs, New York, in late Cambrian times, and resemble still-living forms in West Australia. Living algae range from pond scums to the giant kelps of the mighty ocean. Blue-green and red algal weeds sheltered trilobites in the Paleozoic. Green algae, which live in both fresh and salt water, make food by photo-

Few Pre-Cambrian fossils are distinctive enough to attract attention. This one, of calcareous algae, formed layered cauliflowerlike masses common in the northern Rockies and in Michigan. *Collenia undosa;* shore of Lake Superior, Michigan. (Photo Michigan Conservation Department)

synthesis and in the process they liberate oxygen. Paleoecologists suspect that they were the agents responsible for creating an atmosphere containing enough oxygen for the first primitive land animals to breathe. In this way, they may have had a fundamental influence on the nature and direction of evolution.

Yeasts, mushrooms, and molds, collectively known as fungi, are also thallophytes, but of so fugitive a structure that the fossil record contains little record of them. Fungi do not synthesize food; they live as parasites on other organisms or feed on organic materials. Lichens are an association of fungi and algae, a mutual benefit association known as symbiosis.

Fungi reproduce by forming spores, which are single cells (seeds have thousands of cells) that have the power to grow into a new plant. Some algae also reproduce by spores; others alternate generations of spore or asexual reproduction with sexual reproduction, like the mosses and ferns.

Bryophytes

The next division in the scale of plant evolution is the bryophytes, which include the liverworts, and the mosses. While these lack vascular tissue (the system of ducts through which liquids circulate), they were the first plants to develop stems and leaflike structures. Possibly they were the first plants to abandon the comfortable ocean environment and creep up moist valleys near the shores where there was a bit of sand and weathered rock in a rocky world.

Freshwater lakes may have provided a marshy shoreline where the ancestor of land plants, combining characteristics of algae and mosses, could develop roots to reach down for moisture, woody stem tissue to hold the plant upright, and a thicker skin to protect it from desiccation in the air. In this environment, sexual generation was aided by wet seasons, and asexual spores were produced which were capable of surviving times of drought. The atmosphere by this time must have been composed of a proper combination of gases to shield life on land from the lethal ultraviolet radiation of outer space.

Primitive land plants are first found as fossils in Silurian rocks. Examples from Australia, Scotland, and North America are rootless, prickly stems tipped by spore-bearing capsules. In the talus below Beartooth Butte near the Silver Gate entrance to Yellowstone National Park, collectors hunt for fossils of a shrublike plant with a stem covered with fine needles that were perhaps rudimentary leaves. In Australia, fossils of a plant with needlelike leaves may represent ancestors of the present-day club mosses.

True Ferns

With dry land a new world to conquer, the already diversifying Silurian plants developed vigorously during the Devonian period. From the lowly pioneers developed several major groups that would dominate the lush jungles of the next periods, the coal-forming Mississippian and Pennsylvanian periods. After 100 million years of evolution, plants covered the world during the Devonian period. These included the filicales, or true ferns, which were the first plants to have branching stems and large complex fronds or leaves. They could do so because by this time plants had evolved vascular systems to conduct liquids from the roots to the leaves, and stems strengthened with lignin.

True ferns are large plants that reproduce by spores formed on the underside of the fronds. From the spores grow small plants which bring together male and female cells to form a zygote, which takes root and grows into a fern. Other members of the group known as the pteridophytes—the sphenophyllums, the horsetails, or calamites, and the club mosses, or lycopods—reproduce in similar fashion.

At Gilboa, New York, a standing forest of rich Devonian vegetation was unearthed many years ago and was reconstructed in the New York State Museum at Albany. It includes *Aneurophyton*, a palmlike fern thirty feet tall, slender lycopods *(Lepidosigillaria)* which would become the ruling class of Pennsylvanian forests, and primitive calamites.

Two members of the lycopods, the *Lepidodendrons* or scale trees, and

Asterotheca miltoni, a typical fern of the Pennsylvanian period. Braidwood, Illinois.

Annularia radiata, whorled foliage of the *Calamites* tree. Stub of stem shows in center of whorl. Pennsylvanian; Terre Haute, Indiana.

an allied form, *Sigillaria,* dominated the Coal Age forests of the Mississippian and Pennsylvanian periods. Scale trees were tall woody cylinders with centers of pith and corky bark marked with rows of scars where the long straplike leaves had dropped off. In *Lepidodendron* the scars form a spiral pattern; in *Sigillaria* they march up the trunk in vertical rows. The former branched like an elm tree, and the upper limbs were festooned with long droopy leaves growing out the branches. *Sigillaria* reached a height of 100 feet but did not branch. It resembled a huge bottle brush.

Calamites, which was smaller than these companion trees, grew in thickets like bamboo. Like its descendants, the modern horsetails, it had a jointed trunk decorated at the joints with radiating limbs. *Sphenophyllum* was an ubiquitous vinelike shrub easily identifiable as a fossil by the wedge-shaped leaves. Growing with these giants of the forest were tree ferns topped by immense fronds of lacy leaves, quite similar to modern-day tree ferns of the southern hemisphere.

Bark of the *Sigillaria* tree. *Sigillaria* was similar to *Lepidodendron* and is often found with it. The leaf scars occur in definite furrows. Pennsylvanian; Michigan. (Photo Michigan Conservation Department)

These ambitious new residents of dry land, encouraged by the mild and uniform climate, spread nearly everywhere over the world. Though they appear exotic to us today, they lacked the variety of modern forests be- cause they had no flowers or color to relieve the unbroken landscape of

Bark from *Calamites*, a tree with stems rarely thicker than a man's arm. These trees, which had pithy centers, are usually found crushed flat. The branches were attached at the joints. Pennsylvanian; Pella, Iowa.

Silicified palm wood, easily recognizable by the "eyes" in the wood. Texas.

green. Even the uplands, though less lush, must have been monotonous without flowering plants or a carpet of grass. There were no bird voices to break the stillness—only the lapping of waves, the rustling of leaves in the wind, and the occasional fall of a superannuated tree.

These fast-growing, shallow-rooted trees flourished in the even climate, which was not hot and steamy as supposed by many, but continuously cool as suggested by the lack of growth rings in the trees and the excellent preservation of plant fossils. Trees fell and were buried in the mud deposited in river deltas and by invading seas. The mud preserved them from decay, and they left their record as the thick coal seams of Pennsylvanian times.

Seed Ferns

Growing beside these plants were others that were preparing for the future. Some ferns had by this time developed the habit of producing true seeds. A seed differs from a spore. It is an embryonic new plant enclosed in a protective case that also encloses enough food to nourish it until it is

Neuropteris scheuchzeri, the thick, fleshy leaf of a seed fern. Mazon Creek, near Morris, Illinois.

established as a new plant. With seeds, plants could now live anywhere that conditions were otherwise favorable. Alongside true ferns grew these seed ferns, which in appearance were so much like the true ferns that differentiation of fossils is often impossible. With them grew another rising class, the *Cordaites* trees, which had slender softwood trunks and long straplike leaves.

The seed ferns appear to have been the ancestors of the cycads and cycadeoides, and *Cordaites* is probably the ancestor of the ginkgo and the conifers. *Cordaites* bore male and female catkins that appear to be a development midway between the fronds of the seed ferns and the cones of the pines and firs. Coal-ball peels from mines in Lawrence County, Illinois, disclose that cycadlike structures existed in the Pennsylvanian period, and this has been confirmed by other late Pennsylvanian and Permian evidence. Plant fossils of the Coal Age are exceedingly common at almost every Pennsylvanian-age coal mine in the world, from Iowa to Pennsylvania, England to South Africa, and even in Arctic regions.

Spermatophytes

The plant kingdom became increasingly complex with the development of the spermatophytes. This division comprises the gymnosperms ("naked seed plants") such as the conifers, and the angiosperms ("covered seed plants"), a class that includes most trees, shrubs, and other plants. The angiosperms have true flowers and depend on the color and scent of their blooms to attract the insects and birds that unwittingly carry pollen from one plant to the pistil of another for cross-fertilization. Seeds develop from the pistil. The pines and palms are the major gymnosperms that flourish today.

GYMNOSPERMS
In the final period of the Paleozoic era, the Permian, the forests began to be taken over by the cycads and other cycadlike trees, with short, barrel-shaped trunks and palmlike leaves—and by conifers and ginkgos. The Pennsylvanian swamps dried up as the climate changed, and the water-dependent amphibians adapted to the change by evolving into reptiles. The most spectacular reptiles, the dinosaurs, would feed and fight in the next era, the Mesozoic. Like the Paleozoic forests, the dinosaurs, too, would disappear.

Early in the Mesozoic Era, in the Triassic period, logs washed into shallow lakes where they were buried by volcanic ash, and became petrified and agatized to form the wonders of the Petrified Forest National Park

Few fossil sites are so widely known as the Petrified Forest National Park in Arizona. Thousands of acres are strewn with logs of coniferous trees, now preserved as colorful red, brown, yellow, and purple agate and jasper. No collecting is allowed in the Park, but several ranches nearby are open to collecting for a fee.

in Arizona. These colorful logs are mostly gymnosperms known as *Araucarioxylon*, distantly related to the modern pines. Their nearest relative today, however, is the grotesque monkey puzzle tree. The cycad fossils eagerly collected in western South Dakota once grew in Mesozoic forests where treeless plains exist today. Such forests spread far across the globe in the Jurassic period, when the evidence from plant life indicates that the world enjoyed more uniformly even temperatures and rainfall than at any other period in its history. This halcyon time has left its organic mark with extensive coal beds in such widely separated places as Alaska, Australia, China, Siberia, and Greenland.

ANGIOSPERMS
In the Cretaceous period, last of the Mesozoic era, new types of plants better fitted to cope with changing environments were ready to take over from the conifers and cycads. These were the angiosperms, to which group most modern plants belong. As the climate grew cooler and the swamps disappeared, the cycads and conifers were succeeded by trees, shrubs, and

Sequoia langsdorfi, a conifer very similar to the modern sequoia. Cretaceous; Alaska.

flowers familiar today. In the Upper Cretaceous rocks of Kansas are found fossil leaves from magnolia, beech, elm, sycamore, myrtle, oak, and fig trees; and similar leaf fossils are found in Tennessee clays. These trees had vascular and reproductive systems, with well-protected seeds, more readily adaptable to changing conditions and better equipped to specialize than the cycads and conifers. This was the time when the redwood and its relative, the metasequoia, grew over much of the present United States.

Perhaps the development of the angiosperms had something to do with the dramatic increase in the number of insects in the Cretaceous. They set up a working partnership—the flowers supplying the insects with food and the insects supplying the means of fertilizing them—a partnership of common advantage that persists to this day.

Few fossils of annual plants have survived from the Cretaceous, because conditions were unfavorable for preservation: the plants lived short lives, and their soft tissues decayed rapidly, leaving little record in the rocks.

Cretaceous vegetation established itself more firmly in the Cenozoic era. New times tested its adaptability. The nature of the growing plants, how large they grew, and the texture of their leaves are sensitive thermometers of climatic change. In the Eocene epoch of the early Cenozoic era, a tropical forest grew in Oregon. The size of the trees, the poorly developed growth rings, and the thick fleshy leaves all indicate that this was a time of year-round warmth and abundant rain. By mid-Cenozoic, in the Miocene epoch, the leaves were thinner and the forest was smaller. Growing conditions had deteriorated.

In the Eocene era, maples, walnuts, and the metasequoia flourished as far north as Alaska and Greenland, but by Miocene times such forests were retreating all over the world. Plants abandoned the far northern and southern latitudes and moved to warmer regions.

Thirty million years ago volcanoes sprayed millions of tons of fine volcanic ash over certain lakes near Colorado Springs, especially near what is now the town of Florissant, Colorado. This ash, a perfect mold material, preserved delicately etched fossils of 250 varieties of leaves—including redwood, cedar, beech, willow, maple, elm, oak, and pine. Scores of species of insects and shells are found in this Oligocene deposit, as well as feathers from many species of birds. Similar fossils are found in the John Day basin of Oregon and in the Miocene deposits near Spokane, Washington. The John Day deposits are famous for the profusion of fossil seeds—such as walnuts—preserved in tough volcanic rock. Hundreds of species of seeds and nuts, completely replaced by pyrite, have been recovered from Eocene clay beds on the isle of Sheppey in England.

Generally speaking, plant fossils are best preserved in the lowlands, where sediments and volcanic ash formed suitable media for fossilization, and where there was less erosion to erase the fossil record than in the

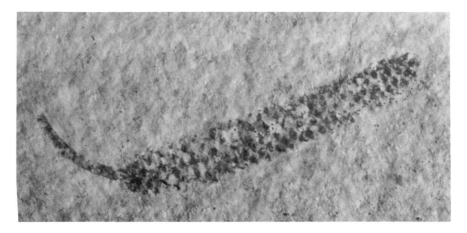

Flowers and fruit are rarely fossilized. This catkin was found along with fish in Eocene shales near Kemmerer, Wyoming.

uplands. Most of our knowledge of the plants of the past, for this reason, is derived from fossils of common lowland trees with firmly textured leaves and woody trunks, branches, and roots. This is the material most commonly available to the collector, too.

A few fossil woods, notably the wood of the sequoia, contain so much tannin that they have resisted decay, remaining as the original wood for millions of years.

ANIMALS

Like plants, animals are classified on the basis of anatomical characteristics that developed from the earliest forms. At their most primitive, these consisted of nothing more than a nucleus and a flowing tissue known as cytoplasm. Animals turned this into increasingly complex nutritional, excretory, sensory, perceptive, and locomotor structures which eventually enabled them to leave the oceans and live on land.

The most primitive animals are so undifferentiated that it is sometimes impossible to tell whether a one-celled organism is plant or animal, or maybe both. Some flagellates, organisms visible only through the microscope, create their food by photosynthesis, like plants, when they are near the surface of the ocean and in the sunlight. But when they are in the darkness of night or of the abyss they feed on organic matter like animals.

Protists

One-celled plants and animals are known collectively as protists. In their various guises they outnumber all other forms of life: the oceans teem with them, and their fossils fingerprint nearly all sedimentary rocks. Some protists build shells or cases of silica, or of magnesium or calcium salts. Some do not. Marine animals subsist on their stored-up fats, oils, and starches. These nutrients from the countless billions of fossil protists, stored in the rocks, probably constitute the raw material of the petroleum that drives our cars and heats our homes.

The shell-less protists include such plants as bacteria, algae (both green and blue-green), and some of the flagellates. Diatoms, which form siliceous skeletons, are plants. Abundant in cool seas, lakes, and streams, their shells form much of the bottom sediment.

Coccoliths, the extinct starlike discoasters, and some protozoa build calcareous houses for themselves, while the radiolarians and silicoflagellates, like the diatoms, prefer siliceous walls. Foraminifera, most abundant and significant of the one-celled organisms, generally build limy walls. Like the others, they are animals.

These tiny creatures have been discussed extensively in a previous chapter. Most of them have been of interest only to geologists and biologists until recent years when amateur collectors learned of the fascinations of the microscopic world. But these tiny organisms occasionally become evident en masse. The dinoflagellates, for example, sometimes create phosphorescent areas or the poisonous "red tide" in ocean waters. Coccoliths, so tiny that powerful microscopes are needed to study them, cause a strange milkiness in the Norwegian fjords on summer days.

The foraminiferan buliminids built the famous white cliffs of Dover with trillions of their fossil skeletons, and the nummulites of the same group of protozoa gave millions of their lens-shaped bodies to form the rocks from which the Egyptians built the Pyramids.

Paleontologists estimate that $2\frac{1}{2}$ percent of all known animal species belong to the order of the Foraminifera, and that 88 percent of these species are known only as fossils.

Like the ferns, foraminifera have an irregularly alternating system of reproduction. Small tests bud off cells that develop into the larger form. These in turn produce a sexual cycle of cells that conjugate with cells from other individuals to produce the small form again.

Sponges are never showy fossils, and with a few exceptions, are never common. *Astylospongia praemorsa;* Silurian; Pegram, Tennessee.

Porifera

Sponges, classified in the phylum Porifera ("pore bearers"), take a long step beyond the one-celled protozoa. Some species of this phylum grow as individuals; some grow in colonies. All live in water, and most of them in shallow seas. Sponges are multicellular creatures, but they lack definite tissue or internal organs except for a digestive cavity. They have no means of locomotion and live attached to a rock or other support. Sponges can be thought of as a fraternity of slightly specialized single-celled animals clustered together into a hollow ball pierced by many holes. Some outer cells have appendages which wave back and forth to force a current of water into the central cavity. Cells lining the cavity absorb food and pass it to the other cells by osmosis. If a sponge were cut up into thousands of

Solitary corals, which are commonly called horn corals, contained a single living animal. They range in size from a flattened button (lower left) to elongated forms. The shape and the prominent radiating septae make recognition easy, but individuals require sectioning for complete identification.

tiny pieces and tossed back into the sea, each part would grow into a new sponge.

Most sponges are supported by a spiny internal skeleton of calcareous, siliceous, or horny material. Soft sponges, such as the familiar bath sponge, occur in tropical seas. Sponges do not attract the hungry marine predator: the spines of the skeleton are unpalatable, and many sponges secrete a repellent odor. Despite this, sponges have been found in the stomachs of large cephalopods in Mississippian rocks in Arkansas. Worms, shrimp, and little fish, however, are not fussy; they shelter themselves in folds and cavities of the sponge.

Sponges have left their record in the rocks since Pre-Cambrian days. Spicules, the sharp spines of the skeleton, often are the only hard parts capable of making a fossil record; the softer sponges are poorly represented.

Classification is based on the form of the skeleton, usually from clues given by spicules. Some fossil sponges are dish-shaped; some are octagonal like *Prismodictya*, some like the stacked chips of *Titusvillia*, or the mushroom of *Coeloptychium*. Delicate glass sponges, so named because of their

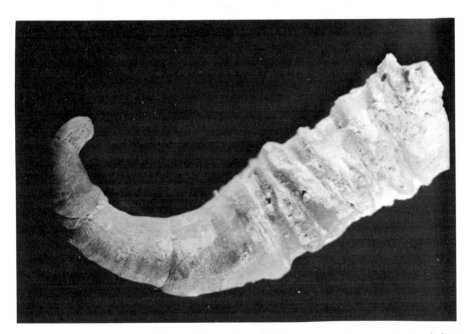

This horn coral grew upright until increasing weight or energetic currents toppled it on its side. Its new growth curved to give the animal an upright position once again. This happened several times before the coral could straighten itself out. The ridges probably reflect periods of active growth under favorable conditions.

fragile, glass-clear skeleton of quartz spicules, swarmed in late Devonian seas. Beautifully preserved sponges occur among the notable examples of marine life fossilized in the Cambrian Burgess shale at Field, British Columbia.

The "sunflower corals," or *Receptaculitids*, found in the Ordovician rocks of the upper Mississippi River valley once were believed to be related to the sponges, but the best opinion now is that they are algae.

Coelenterates

For the collector the phylum Coelenterata ("hollow gut"), which is placed next above the sponges in the scheme of life, affords more rewarding trophies than those previously discussed. Sponges are rarely preserved well enough to hold a place of honor in the cabinet. But the corals, which with jellyfish and sea anemones are classified as coelenterates, occur in

Commonly called organ-pipe coral, *Syringopora* is a widespread Silurian fossil. (Photo Michigan Conservation Department)

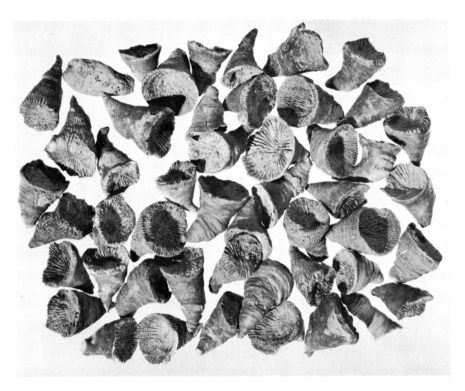

Typical horn corals that a collector might find weathering from Paleozoic shales. Several species are represented, mostly *Zaphrentis* and *Heterophrentis*. Devonian; Michigan. (Photo Michigan Conservation Department)

abundance and great variety. Some, such as the Petoskey stone, are so well preserved that they can be used for ornaments and jewelry.

Coelenterates have the distinction of being the first animals to possess well-developed tissues; they have a body wall formed of two layers of specialized cells. In most species the body cavity is divided by radial partitions, as in the horn corals. Corals evolved from an animal like the simple, still-existing hydra. This is little more than a tube fringed at one end with tentacles that bring food into the gut and a foot at the other end to anchor the hydra to a stone. In the rugose or horn corals, folds of tissue built radial septa (partitions) inside the body cavity while the other end grew a limy horn to hold the organism upright on the muddy bottom. Fossils consist of the horn and septa that supported the soft body. Rugose corals are found in rocks as old as the Ordovician period; later they evolved flat forms that could rest on the mud or grew in masses or as

Typical of the colonial corals is *Hexagonaria*, a Devonian species from Michigan. Each cell held a living animal. This is the familiar Petoskey stone.

individuals packed together for common support. Such forms reached their peak in the Devonian period.

Colonial corals resulted from asexual budding, which created a mass of descendants of one individual. These include the hexacorals, so named because the septa are organized on a basic unit of six, although the individuals, called corallites, may be hexagonal or round. Such fossils have a pockmarked surface. The familiar brain coral belongs in this group. Among hexacorals, which have been known since Triassic times, are the Miocene corals of Tampa Bay, Florida.

Paleozoic tabulate corals include a number of familiar species. These colonial forms are represented by *Halysites* and *Catenipora*, which are identified by their chainlike patterns; the Devonian *Hexagonaria*, which is the familiar Petoskey stone; and *Favosites*, which resembles a honeycomb. One living coral that has no superficial resemblance to other corals is the sea fan. This horny-tissued favorite of beachcombers is related to the precious Mediterranean red coral.

More primitive coelenterates occasionally appear as fossils. Examples

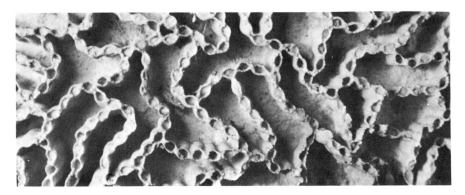

Halysites, the chain coral, is another colonial coral. The polyps lived inside the links of the chain. Silurian; Wisconsin.

are the millepores, which were important reef builders with their massive colonies pierced by tubes, and the stromatoporids of Paleozoic age, which built small reefs of shapeless lumps or sheetlike layers. Jellyfish lack the makings for fossils; their tissues are too soft, but impressions of their tentacled bodies occasionally appear in rocks as old as the Cambrian period.

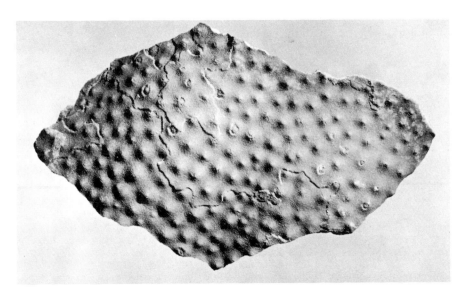

Top view of a stromatoporid coral, *Parallelopora*. These corals have extremely fine pores compared with other corals and in small specimens may be mistaken for bryozoans. Devonian; Michigan. (Photo Michigan Conservation Department)

Bryozoans

The bryozoans, or moss animals, which occupy the next niche upward among the phyla, masquerade as stunted relations of the corals; but they did add something to the complexities of life. Unlike the corals, they developed a body cavity differentiated from the gut, specialized organs to clean themselves of foreign particles and for reproduction, and a nervous system capable of making the muscles contract.

Byrozoans, of which 3000 species live today compared with 4,600 species known only as fossils, grow by budding into colonies as much as a foot across, though most are mere twigs. Most lived near shore in shallow seas. Many modern species make up a good part of the "seaweed" that grows on wooden pilings. The various species are identified by the shapes of the tubes in which they live and by characteristics of the colonies.

One of the most striking fossils is *Fenestrellina*, the lacy bryozoan that

Some typical bryozoans. Upper left and bottom center are irregular massive types. Far right and lower left are the branching, sticklike types. Upper center is a common type that encrusts other fossils. Lower right is a thin, fanlike type that resembles coarse cheesecloth. Bryozoans resemble corals, but their individual chambers are much smaller.

Archimedes is named after the Greek who invented the screw. The "screw" is the axis of a much larger animal; attached to the screw are wide, lacy fronds. This bryozoan is a common Pennsylvanian and Mississippian fossil in the United States. A typical screw is shown below the rarer frond shown above it. The hole in the center of the frond is where the screw fits. Mississippian; Attica, Indiana.

looks like a fan and is to be found in Devonian to Permian rocks. Another favorite among collectors is *Archimedes*, the screwlike central axis from which fronds like those of *Fenestrellina* projected. Usually the delicate fronds are missing from the robust central support. This fossil is confined to Mississippian-age rocks.

Ordovician rocks contain rounded colonies of such major stony forms as *Dekayella* and *Prasopora;* the Cincinnati area is particularly rich in bryozoans. Bryozoans became slender and lacy in the late Paleozoic; then a great outburst of evolutionary energy in the Mesozoic and Cenozoic developed a host of complex and striking forms. Bryozoans are easily mistaken for corals, but there are major differences visible in the laboratory. Bryozoan individuals are commonly smaller and more delicate than corals. Individual pores in a bryozoan colony are smaller than a pinhead; the

Modern and fossil brachiopods. Upper left is *Tropidoleptus carinatus*; Devonian; Griegsville, New York. Lower left is a modern *Terebratula*, similar in shape to *Tropidoleptus* but not closely related. Upper right is *Lingula*; Ordovician; Elgin, Iowa. Lower right is modern *Lingula*. The genus *Lingula* has flourished for 400 million years.

pores of corals are larger. Bryozoans from Paleozoic rocks are often found as a thick film or crust on other fossils, but corals are not. Proper identification requires sectioning to disclose the fine internal structure of the colony.

Brachiopods

Animals classified as the Brachiopoda ("arm foot"), although a superficial acquaintance would not indicate it, have much the same plan of life as the bryozoans. Larval bryozoans start with a shell covering but discard it as they develop. Brachiopods keep the shell, grow larger than bryozoans, and discard the colonial way of life. Like the latter, they have existed since the Cambrian period and have diversified into thousands of species, of which only 200 are living forms. One of these, the small, smooth-shelled *Lingula*, lives today in muddy sea bottoms just as it did 400 million years ago. Presumably, ancient brachiopods, like those today, inhabited shallow seas, anchored to the bottom by a stemlike pedicle

Representative brachiopods of the major types. Upper left is *Meristella haskinsi;* Devonian; New York. Upper middle is *Rhipidomella leucosia;* Devonian; New York. Upper right is *Leptaena;* Silurian; Tennessee. Lower right is *Stropheodonta demissa;* Devonian; Ohio. Lower left is *Brachyspirifer angustus;* Devonian; New York.

cemented to a rock or buried in the sediment. For a passive animal such as this, the shell is protection against predators and stormy waters.

Brachiopods, which are also called lamp shells, are bivalves, but the two shells are not alike in size or shape. One shell usually overhangs the other at the hinge, so that when viewed from the side the difference in shape is apparent. From the front, brachiopod shells have bilateral symmetry, that is, a line down the center of the shell will divide it into two halves that are mirror images of each other, like a human face.

One shell may be flat, the other concave or convex, or both may be convex in different degrees. The shells may be shaped like wings, as in the spirifers, or they may be smooth, or grooved across or up and down; or they may even be ornately spiny. One group of brachiopods, the inarticulates, holds the shells together by ligaments and muscles; but the articulate types are additionally joined by a tooth or teeth that fit into a socket in the other shell. This forms the hinge on which the shells open and shut. Brachiopods are small, usually an inch long or less; a three-inch specimen is a giant. Clams, which at first glance resemble brachiopods, are generally much larger and generally grow two equal shells.

Collectors not only have pseudofossils to contend with but also some fossils that masquerade as others. At first glance these broken fossils appear to be hickory nuts, but closer study shows that they are *Atrypa,* a common brachiopod of the Devonian period. (Photo Michigan Conservation Department)

Brachiopod fossils are exceedingly abundant in Paleozoic rocks, scarce in the Mesozoic. To the amateur collector they are both a joy and a problem. They are common and easily collected; they exist in great variety, and many of them are beautiful. But identification is often difficult or even impossible without tedious study of such minute distinctions as texture of the shells and muscle scars. Some species from unrelated stocks developed outwardly similar shapes, and for this reason they cause great confusion.

The brachiopods, particularly the more specialized articulates, reached their zenith in Ordovician seas. The Devonian oval terebratulids are well represented in the rocks of New York and the Midwest, along with the large plump shells of the pentamerids. Mississippian shales are the place to look for the strongly striated, sharply beaked rhynchonellids. The nearly flat Ordovician strophomenids, with their wide hinge line and fan shape modeled by the familiar *Rafinesquina*, contrast markedly with the spectacular spirifers of Ordovician to Permian times. Waldron, in Decatur County, Indiana, and Pegram, Tennessee, have yielded excellent specimens of Silurian spirifers. Erie County, New York, is famous for Devonian spirifers, and northern New Mexico for giant Pennsylvanian ones. The Medusa quarry near Toledo, Ohio, is notable for large, pyritized Devonian brachiopods found associated with trilobites. Brachiopods today are represented by the terebratulids and rhynchonellids as well as a few more primitive, long-lived forms.

Mollusks

Far better equipped for the long pull were the members of the next phylum, the Mollusca ("soft bodied"), an even vaster group of shelled organisms. The mollusks, second largest living phylum in number of species after the arthropods, include marine, freshwater, and land organisms. Familiar members are the snails, clams, and cephalopods. Like the brachiopods, some burrow in the bottom or anchor themselves to a rock, but others walk or swim, despite the handicap of dragging around a primitive skeleton poorly engineered for locomotion of any kind.

Five basic classes are recognized among the mollusks. The least advanced are the chitons, which live in armor formed of eight transverse articulated segments. The scaphopods, or tusk shells, are long and slender like an elephant's tusk. The gastropods, or snails, form the largest single group, followed by the pelecypods, or clams, which hide themselves between two shells. The last order, the cephalopods, includes the largest and most intelligent as well as the most evolutionarily versatile of the invertebrates. Modern cephalopods—particularly the octopus and giant squid—are repulsive and even dangerous to man.

CHITONS

Chitons occur in the fossil record since Paleozoic times. Even though their shells are hard enough to leave a mark in the rocks, chiton fossils are not plentiful. Living examples, which are numerous on the Pacific coast, cling to a rock with their lower surface, which is an unprotected fleshy foot, and feed by scraping algae from the rocks with a rasplike tongue.

SCAPHOPODS

Tusk shells sit partly buried at an angle in the mud of the sea bottom. The class has never included many species, although it goes back as far as the Silurian period. Their little shells, rarely more than an inch or two long, are easily identified, although they may be confused with worm tubes.

GASTROPODS

The Gastropoda ("stomach foot"), or snails, have recognized few limitations. They are the only invertebrates except the arthropods (insects, arachnids, etc.) that have learned to live on dry land. They did this by turning the mantle used by marine invertebrates to absorb oxygen from the water into a "lung." They also learned to do without salt water and live in fresh water. As a result, they can be found almost everywhere on the surface of the earth. They far outnumber the other mollusks, with some 80,000 living species and thousands of fossil species dating back to early Cambrian rocks. The abalone, which has a pretty shell and delicious meat, is a snail. So is the ugly garden slug, which has a tiny vestigial shell buried in its back.

The common snail moves on a muscular foot while carrying its protective shell piggyback. Most snails also have a hard, horny trapdoor (called an operculum) on the foot. This neatly closes the aperture, like a manhole cover, when the animal retreats into its shell. The snail is guided by projecting tentacles. In some species these tentacles are equipped with eyes. Many snails are scavengers or vegetarians, but others rasp holes with their tongues through the shells of mollusks and draw out the succulent contents.

Gastropod shells are as varied as their habits and environments. Some are cones, some tubes. Most forms have coiled shells, either a flat coil or a spiral cone. Early Paleozoic forms were caplike or flat-coiled shells living in shallow water or on mud flats.

The class moved like a conquering army into the Ordovician period. Shells became more complex, such as the spiral cone of *Trochonema*, which is like the modern trochus, or top shell. By the Silurian period high-spired forms had appeared, and by the late Paleozoic such shells as *Shansiella* and *Pleurotomaria* rivaled the cephalopods for beauty and complexity.

Land and freshwater forms date from Pennsylvanian times. Land snails

Gastropods, which are snails, always show some kind of coiling, a basic shape they have retained for 500 million years. Upper left is an internal cast of *Turritella mortoni;* Eocene; Virginia. Upper right is *Straparolus;* Pennsylvanian; Nebraska. Lower left is *Trepospira depressa;* Pennsylvanian; Oklahoma. Lower left center is *Euphemites,* in which the coils are covered by the last part of the shell; Pennsylvanian; Oklahoma. Lower right center is an internal cast; Silurian; Illinois. Such a cast bears little resemblance to the outer shell and is difficult to identify. Lower right is *Platyceras bucculentum,* often found in the calyx of crinoids; Devonian; Ohio.

are thin-shelled and relatively small and are rarely found as fossils, for they died without being entombed in protective sediments.

Snails have left not only their bodies but also their trails in the fossil record. They have also left their grim calling card—a neatly drilled hole— in the fossil shells of ancient clams, brachiopods, and other snails. Occasionally one is found neatly coiled on the calyx of a crinoid, where it fed on the excreta of the host. Such specimens, which usually are the gastropods *Platyceras* or *Cyclomena,* are rare and desirable additions to a collection.

PELECYPODS

The Pelecypoda ("hatchet foot") differ from all other mollusks in one particular: they have two valves or shells, usually composed of a calcareous material, joined by a toothed hinge. In Europe they are termed lamellibranchs by paleontologists; the common names for them are clams, oysters and mussels. In the pelecypods, one shell does not overhang the other as in the brachiopods; from the side they are mirror images of each other and meet at an evenly shared hinge line, even though one valve may differ from the other in convexity or shape.

Some pelecypods move slowly but without great effort by protruding a spade-shaped foot, planting it, and pulling themselves forward. A few, such as the acrobatic pectens, swim backward by suddenly flapping their valves together so as to expel water, a crude jet propulsion. Most clams can and do burrow in beach sand or the marine bottom, often with incredible speed. A few, such as the razor clams, can leap as far as a foot

Rudistids look more like horn corals than clams. One valve is reduced to the radiating "cap" and the other valve has developed into a cone, attached by its narrow base. Its large size serves to separate it from horn corals. Cretaceous period.

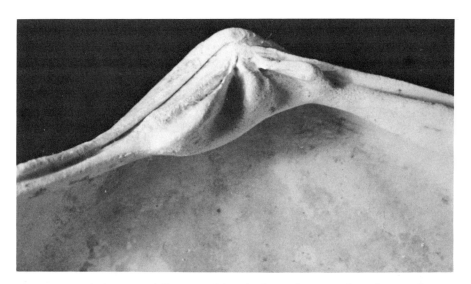

Major types of clams are differentiated by the hinge lines, tooth sockets, and types of teeth. This one is *Tivela stultorum;* Pleistocene; California.

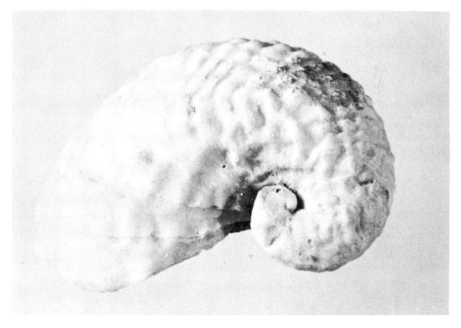

Exogyra cancellata, an oyster. Coiled oysters such as *Exogyra* and *Gryphea* resemble snails, but they differ in that they are bivalves, are larger than most snails, and show growth ridges on their shells. Cretaceous; New Egypt, New Jersey.

Chlamys, a scallop. Though more symmetrical bilaterally than many clams, so that it resembles a brachiopod, its much larger size and the nonsymmetrical hinge line identify it as a clam. Miocene; Maryland.

by a sort of pole-vaulting maneuver. Some even bore holes in wood or rock and become voluntary prisoners there as they grow too large to emerge.

Shipworms, which riddle wood piling and hulls, are a form of pelecypod, not a true worm. The fossil teredo "worm" is such a pelecypod. Oysters moor themselves to a rock; and horse mussels and some others resist the force of the waves by attaching themselves by a hairlike mass of cords, known as a byssus, to solid objects near the tide line. One clam, *Tridacna,* is the largest pelecypod, a giant whose bathtub-size shell often weighs as much as 500 pounds. The stories of venturesome men trapped and drowned by this clam appear to be strictly adventure fiction.

Pelecypods have participated in the evolutionary adventure since early Paleozoic times, but they have persisted in their Paleozoic ways with little change because they have been able to find comfortable environments instead of adapting to changing ones. For this reason, many fossil species closely resemble living marine and freshwater forms. This is a great help in identifying them.

Clams are fairly common in the rocks, often as casts, and some specimens of *Trigonia,* a Cretaceous pelecypod, have been found with the pearly shell still gleaming after 100 million years. *Inoceramus,* a giant, thick-shelled Cretaceous form found in the chalk beds of Kansas, had a shell an inch thick and three feet long.

Fossil species are identified primarily by the teeth on the hinge. For example, the prominent teeth and sockets of the schizodonts lock tightly; the unios (freshwater mussels) which developed in the Triassic period are classified with them. The dysodonts gather together a variety of almost toothless clams, including the important pectens (such as the modern

Uranoceras, a nautiloid, displays the simple straight sutures characteristic of the cephalopods before the ammonites appeared. McCook, Illinois. (Photo Field Museum of Natural History)

scallop); *Mytilus*, the saltwater mussel; and the oysters *Exogyra* and *Gryphea*, which emulate the gastropods by making an effort to coil. A fourth group, the heterodonts, includes shells related to the modern *Lucina*, *Astarte*, and the tellins. These have an unequal number of lateral teeth on the valves. Boring and burrowing clams, such as ark and razor clams, are classified in the group known as desmodonts.

One form of Cretaceous clam is likely to confuse the collector. This is the group of pelecypods known as the rudistids, in which the lower valve matures into a ribbed cone resembling a horn coral, while the upper one is reduced to a flat toothed lid.

CEPHALOPODS

The Cephalopoda ("head foot") have left their stuck-in-the-mud relatives far behind. They are not only smarter and much more highly developed, but they are also willing to live dangerously. Some had the good sense to discard the shell and depend on speed, excellent sight, and sheer meanness for protection. Exclusively marine, they are free-swimming and eat other marine animals.

The cephalopod's highly developed head contains eyes and a horny beak with jaws. The head is surrounded by muscular arms called tentacles.

Small ammonites resemble snails, but ammonites will always show tracery of sutures on the surface of the shell and matching divisions on the inside. Snails have only one chamber to the shell and usually are not ornamented. Left is an internal cast of a snail from the Silurian of Bohemia. Right is *Dimorphoplites*, a small ammonite from the Cretaceous; Folkestone, England.

Beneath the head is a tube formed of a fold of skin through which the animal can squirt a jet of water to propel it either forward or backward. Some fossil cephalopods were as much as fifteen feet long, and the living deep-sea squid is more than twice as large, large enough to attack even whales. It is the largest living invertebrate.

Nautilus is the only living representative of the hundreds of species of nautiloid cephalopods of ancient seas. It builds its spiral shell as it grows, periodically adding a new living chamber at the end and sealing off gas-filled chambers behind it to give it buoyancy for fast and effortless swimming. It floats with the coil up and the head and tentacle parts horizontal. This device, which it developed in the Paleozoic era, marked it off from some primitive straight forms, such as *Volborthella*, and loosely coiled species that probably crawled on the bottom.

By late in the Paleozoic, the nautiloids became as ornamented as dowagers, and, like the dowagers, began to disappear as the next era wore on.

Their place was taken by a type bursting with evolutionary energy and a taste for variation. These were the ammonoids—coiled forms that would seem to have set elaboration as their goal. The goniatites, a simple type,

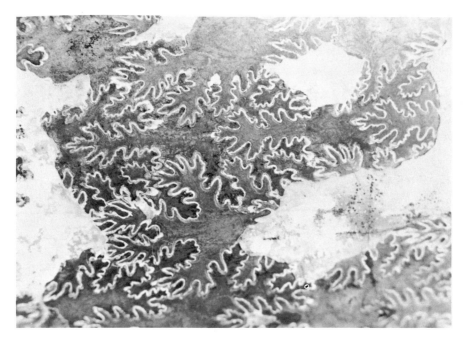

Detail of shell of *Baculites*, an ammonite that did not coil. The sutures are usually visible only if the pearly outer shell is removed. Cretaceous; South Dakota.

Ammonites flourished in the Mesozoic era, then became extinct. Species differed in the intricate suture lines, and these are the major means of identification. Ammonites ranged in size from a fraction of an inch to four feet in diameter. *Harpoceras;* Jurassic; Somerset, England.

are common Mississippian fossils. Their straight sutures (the "joints" between chambers) evolved in later forms into folded and crinkled sutures that left their intricate tracery on the surface of the fossils of the Mesozoic ammonites and ceratites, Paleontologists looking for a function for this elaboration have theorized that the fanciful sutures gave the shell strength to resist the pressure of deep water. But this is pure speculation. At the same time some straight uncoiled cephalopods such as *Baculites* were leaving their pearly remains in the rocks of Cretaceous sea bottoms.

By Cretaceous times, the ammonites had grown huge, up to ten feet in diameter, and were conspicuous for the knobs and suture patterns that marked them as the dandies of the sea. Hundreds of species of these ornate fossils have been found in Europe, where they are an important index fossil. As if worn out from their evolutionary zeal, they became extinct at the end of the Cretaceous period.

Ammonite fossils were the ones that led William Smith to develop his law of correlation, and study of these fossils caused other Europeans in

Ammonites and other cephalopods with external shells grew by adding a new living chamber periodically. They then moved out of the old, sealed it off, and moved into the new chamber. These old chambers are visible if the ammonite shell is cut in two, as this one was.

the nineteenth century to expand this law into correlation by zones and stages. Fossils of ammonites and other cephalopods are widely distributed and abundant; they make clearly recognizable fossils and show rapid evolution. For these reasons they have been regarded as ideal index fossils. They have also exemplified the theory that active animals evolve more rapidly than those that are content with a passive life on the bottom.

In the extinct belemnoids the external shell had been discarded and replaced by an internal, cigar-shaped skeleton that is not uncommon as a fossil in Mesozoic rocks. Related forms include the cuttlefish, or sepia, which is a living species today as well as a fossil. The cuttlefish "bone" used in bird cages is the internal skeleton of this cephalopod. Other shell-less forms, such as the octopus, are uncommon fossils in Mesozoic rocks. Modern representatives of the Coleoida, the class to which these belong, are the coiled *Spirula*, which looks like a tiny hunter's horn, and the paper nautilus.

Belemnites were like modern squid, but had strange, cigar-shaped internal shells such as the one shown here. The shells are common in Cretaceous formations.

Worms

The lowly worm claims a high place in the hierarchy of evolution. Worms, which are animals with a head, sense organs, and a posterior, and which crawl on their bellies, have so thoroughly worked their way into the pattern of life that they have been subdivided into a number of phyla.

But only one cuts much of a figure as fossils, because worms lack hard parts. Instead, they left behind trails where they crawled or burrowed in soft mud, although the nature and origin of these fossil evidences are debatable. Furthermore, most worms have left little evidence of their long evolutionary history. An exception exists in Cambrian rocks, however, where fossils of annelid worms, formed of ringed segments like the modern earthworm, are known from such favored spots as the British Columbian Burgess shales.

Worms move by rhythmic contractions of their segments. Some segments developed into such specialized organs as a head, tentacles, etc. They also have efficient digestive, excretory, circulatory, and nervous systems. Some have eyes.

Many annelid worms are known only from Cambrian and more recent fossils of horny or hard jaws called scolecodonts, which have been discussed in a preceding chapter. These bear toothlike grinding surfaces. A few species of annelid worms are represented by fossils of the limy tubes in which they hid their soft bodies. These tubes are usually found attached to fossil shells.

Grouped with the annelid worms are several fossils that cannot be neatly placed on a step of the evolutionary ladder. These include the conularids, which first appeared in the early Paleozoic and became extinct early in the next era. A typical conularid, such as *Conularia*, has a thin phosphatic shell that takes the shape of an elongated Egyptian pyramid perched on its point and resting on a disc that holds it upright.

Arthropods

The course of destiny turned a decisive corner between the worms and the next phylum—the Arthropoda. This gigantic assemblage of nearly three-quarters of all known animal species includes the insects, the spiders, centipedes, shrimp, lobsters, crabs, barnacles, and trilobites. Arthropods are segmented like annelid worms, but they wear horny, jointed armor that allows the body to move yet supports it like a skeleton. It also protects the tissues from dehydration in the air. This armor is strengthened in certain spots by deposits of carbonates and phosphates. Arthropods molt this exoskeleton periodically as they grow.

Complete trilobites are rare; their bodies were easily broken apart by scavengers or ocean currents. This specimen has started to separate. *Calymene slocomi*, Ordovician, Minooka, Illinois.

Arthropods take their name—"joint legs"—from the paired jointed legs or specialized appendages such as jaws, limbs, claws, or antennae attached to the body. Insects have wings in addition to legs. The typical arthropod has a head formed of several fused segments, a midsection or thorax, and one or more posterior sections. Their evolutionary progress is reflected in the possession of a mouth, digestive, and excretory systems, a heart, brain, and either simple or compound eyes. Some even possess sound-detecting organs. A highly successful system of tubes that take oxygen directly to the tissues has made it possible for insects to inhabit nearly every region of the earth.

TRILOBITES

The sudden appearance of the trilobites, most ancient of the arthropods, in the Cambrian rocks demonstrates dramatically how much evolutionary history must have been lost. Such highly developed organisms must have evolved over millions of years from simpler forms that left little record of their ancestry. Well-advanced brachiopods and mollusks are found with the

The cephalon is usually sufficient to identify a trilobite. The shape, type of eyes, shape of the nose, or glabella, and the ornamentation will identify the species. This is *Dalmanites*, a highly ornamented species. Silurian; Pegram, Tennessee.

Almost every piece of this slate has a trilobite or part of one in it. The locality, Mt. Stephen in British Columbia, is famous for its large Cambrian trilobites.

trilobites, so it is probable that their ancestors, too, had not developed parts durable enough to become fossils.

One small living group of caterpillar-like creatures is believed to be the cross link between the annelid worms and the trilobites, making this group, Onychophora, perhaps the oldest surviving form in the phylum from the environment of 600 million years ago.

Trilobites, distinctive and immensely varied in structure, stand near the top among fossils that collectors prefer. They were the kings of Paleozoic seas, then gradually disappeared with the era. Their name calls attention to the body structure, which consists of three parts or lobes running the length of the body. Transversely, the body is divided into the head or cephalon, the middle or thoracic area made up of articulated segments supported on pairs of legs, and the tail, or pygidium. The cephalon is like a pudgy face, with cheeks and a raised central area known as the glabella. If the trilobite is not a blind variety there also are antennae and eyes. The mouth was on the lower surface of the cephalon so that the animal could feed as it crawled or swam near the bottom.

As the animal grew it molted its chitinous exoskeleton repeatedly and

Complete trilobites are uncommon, especially of a species such as *Phacops*, which molted as it grew. But pieces of the molted armor are common. This is a typical collection of such pieces. Devonian; Sylvania, Ohio. (Photo Michigan Conservation Department)

hid away, like a soft-shelled crab, until the new covering hardened. Fossils often consist of this discarded skeleton, or pieces of it. Even when the whole animal was fossilized, usually as a cast, the appendages rarely were preserved.

The most primitive trilobites, such as *Olenellus*, have a spiky, rudimentary pygidium. The next step up is represented by the bulbous-faced forms of which *Dalmanites* and *Phacops* are typical. *Dalmanites* has a spike trailing from the rear of its pygidium. All these species are widespread in Silurian and Devonian rocks.

The mechanism by which the trilobite shed its skin evolved, giving another criterion for classification. This is determined by the location of

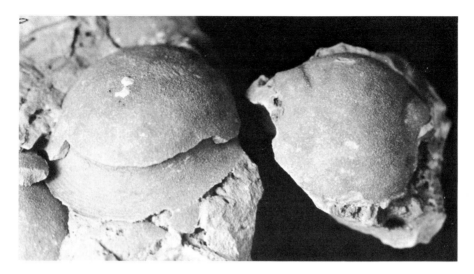

Bumastus, a large trilobite with little ornamentation. These poorly detailed internal casts are typical of preservation in Midwestern dolomites. The tail, or pygidium, is at left, the head, or cephalon, at right. Silurian; Chicago.

Barnacles appear to be related to clams, but they are related instead to trilobites, crabs, and insects. This is *Balanus concavus*; Miocene; Maryland.

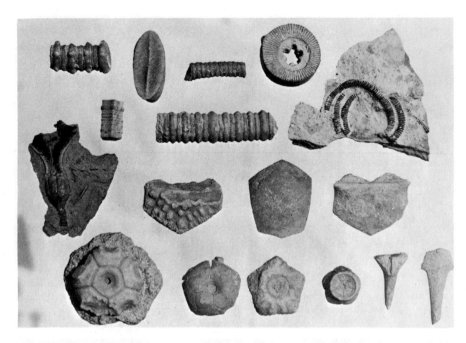

Crinoid stems (columnals) are commonly found as segments or in sections (pictures at top). Most are round, some are hexagonal or elliptical. Less commonly found are the basal plates, five- or six-sided, smooth or bumpy, such as the individual ones (center) or ones joined together (bottom row). A few species have spines (lower right). A complete crown with stem, calyx, and feathery arms is in center row left.

the creased lines in the cephalon where the hard covering split. In early types, the split came along one of the lines in such a way that the skin, as it peeled away, interfered with the animal's sight. In later forms a better pattern was developed that did away with this inconvenience.

Another group, the opisthoparians, largest of the trilobites, persisted throughout the Paleozoic era. These giants have left fossils thirty inches long. At the other extreme were the blind agnostids, so tiny that twenty adults would find plenty of room on a dime. Some trilobites, such as *Paradoxides*, displayed a huge head shield, *Isotelus* smoothed out its glabella and tail, and its surface lost all visible trace of its ancestral segmented structure. *Bumastus* followed suit.

The trilobites are gone, mileposts on the road of evolution but as dead as the Pharaohs. More familiar as living arthropods are such restaurant fare as lobsters, shrimp, and crayfish, and their companion scavengers of the seas, the water fleas.

Allied to them are the ostracods—minute living and fossil clamlike

Crinoid attachments are uncommon as fossils except for *Eucalyptocrinus*, which has left many fossils of its "roots," especially in the Silurian of the Midwest. Silurian; Waldron, Indiana.

forms mentioned in the chapter on microfossils. Ostracods are abundant in both fresh and salt water and have left their shells in rocks of all periods since the Ordovician. Several other groups are akin to the ostracods, but only the fairy and brine shrimps and water fleas among them are known as fossils.

Strange as it may seem, barnacles *(Cirripeda)*, the curse of the ships of olden days, are true arthropods. Anatomically they are crustaceans, but this is evident only in the free-swimming larvae. The relationship is concealed because barnacles cover themselves with limy plates after settling down on some suitable permanent support where they grow with unparalleled rapidity. Their conelike fossils have been present in rocks since the Silurian period.

Crabs, lobsters, and shrimp of the sea, the crayfish of fresh waters, and the pill bugs of damp spots on land constitute the class of malacostracans. Hard shells from this group, especially the shrimp of the sea, appear as fossils as far back as the Devonian. Crabs and lobsters are a much more recent development.

Horseshoe crabs have changed little in the last 300 million years. This specimen of *Euproöps* resembles the much larger horseshoe crab common today on the Eastern shore of the United States. Pennsylvanian; from a concretion found in Mazon Creek, Illinois.

SCORPIONS AND SPIDERS

The most fascinating fossil arthropods, except for the trilobites, fall into the divisions dominated by the scorpions and insects. Granddaddy here is *Limulus*, the horseshoe crab, which can call fossils from Devonian time its forebears. Like *Limulus*, these have a broad oval body terminated by a spiky tail. They recall the ancient scorpionlike animals known as eurypterids, which had their proudest moments in the Silurian period. Monsters several feet long with front claws shaped like paddles, they have been called the terror of ancient seas. Before becoming extinct they moved into rivers and brackish swamps.

Living reminders of these monstrous ancestors are the spiders and scorpions of today. These have eight legs, which is one easy way to distinguish them from the insects, which have six. Scorpions and spiders, like insects, have tracheae (air passages) to carry oxygen to the tissues. Many build webs.

Scorpions and wingless insects may have been the first air-breathing

Canal construction in New York State in the early nineteenth century exposed splendid fossils of eurypterids, some a foot long. This is *Eurypterus lacustris;* Silurian; New York. (Photo Field Museum of Natural History)

animals and the first to undertake conquest of the land. Later, perhaps, snails and worms followed their lead.

The first fossil record of the scorpions has been found in Silurian rocks, a period in which spiders were also making their debut. Centipedes occur in Pennsylvanian rocks and in the Oligocene amber from the Baltic sea.

INSECTS

But all these other arthropods are exceeded in numbers and in ubiquity by the insects, which are the only invertebrates as well adapted to this earth as man himself, perhaps even better. If man should eventually go the way of the dinosaurs, possibly by being overly clever with nuclear weapons, the insects would appear to be the heirs to his earthly domain. Flying on wings, with an efficient system of getting air to the tissues, light in body weight, and so variable and adaptable that there are perhaps a million species, they have made themselves at home everywhere that man has gone. Their rapid evolution has helped many species to become immune to insecticides and vast doses of radiation in the past thirty years.

Winged insects were preceded by some wingless forms, of which the silverfish is a contemporary example. The winged species made their bow in Pennsylvanian times, although paleontologists expect to find traces of them in Mississipian and even Devonian strata. Coal Age forests heard the hum of giant dragonflies with wingspreads of more than two feet. Contemporary with them were cockroaches that would be ashamed of their puny present-day descendants, and grasshoppers and crickets. The Pennsylvanian period has often been called the age of cockroaches, with some 800 species differing little from their descendants today. As one paleontologist remarked: "Imagine a boarding house in the Pennsylvanian period." The first winged insects were limited by inability to fold their wings; grasshopper-like insects appear to have been among the first to solve this problem.

The next period, the Permian, brought forth the beetles, followed by flies, bees, and wasps in the Jurassic, and moths and butterflies in the early Cenozoic.

Early insects are not common fossils because of the fragility of their bodies, but exceptionally favorable circumstances for preservation existed in some places. Highly prized fossils have been found in Pennsylvanian rocks in coal mines of Illinois and Indiana, in amber from the Cretaceous of New Jersey, in Permian lake beds of central Kansas, in the Oligocene lake deposits of Colorado, and in Miocene nodules of California.

Echinoderms

Animals grouped in the next phylum, the echinoderms, unlike the insects, stayed in the sea. Some are still there: the sea urchins, feather and brittle stars, starfish, sand dollars, sea cucumbers, and crinoids. Some, such as the eagerly sought edrioasteroids, cystoids, and blastoids, no longer exist except as fossils.

These animals, whose phylum name means "spiny skinned," almost live by man's decimal system. They build on a plan of five; most of them have fivefold body symmetry. Their skeletons of calcite, like the skeletons of vertebrates, grow to keep up with the growth of the organism. A well-developed and complex water-vascular system provides pressure to the tube feet and aids in respiration. Digestive, nervous, and reproductive systems are efficiently organized. In arrangement of the body cavity and in some other ways, the echinoderms appear to be related to the most primitive chordates, which are predecessors of the vertebrates and, eventually, man.

Some echinoderms, such as the crinoids, attached themselves to the sea bottom by a fairly stiff stemlike support as much as fifty feet long. Others, including most modern crinoids, are free-swimming. Most of the fixed forms of crinoids, eocrinoids, blastoids, and cystoids are known only as fossils as far back as Cambrian times. Like many other organisms in

Echinoids are not common fossils before the Mesozoic. Urchins and sand dollars, such as those from the Texas Cretaceous shown in the upper row, are like modern examples. The spines and isolated plates are more difficult to identify. The plates resemble crinoid plates with a tubercle in the center where the spines were attached, and the spines have swollen bases and often tiny spines or ridges. The jaw, called Aristotle's lantern, is shown below the large urchin. When found alone it is not easy to associate with echinoids.

Cambrian rocks, they obviously developed from ancestors that have not been preserved, or at least have never been found. Of the fixed forms, only crinoids survived the Paleozoic era; other echinoderms reached their peak after the older forms were gone.

Thousands of fossil crinoids, blastoids, and cystoids are known. They are classified primarily by the structure of the plates that enclose the body, particularly the basal plates. One group, the carpoids that existed in Cambrian to Devonian times, had stems that resemble tails. Another from the early Cambrian, the edrioasteroids, resemble starfish perched on disc-like bodies. They persisted until Pennsylvanian times.

Cystoids and blastoids have some common features. They evolved from primitive cystoids with irregularly placed body plates and indistinct sym-

Blastoids were unsuccessful relatives of crinoids and disappeared before the end of the Paleozoic. They are easy to identify by their five grooves and their small, rounded shape. Most common is *Pentremites*, upper row. *Orbitremites*, at lower left, and *Schizoblastus*, at lower right, are less common. Mississippian; Illinois.

metry into the highly organized blastoids, such as the common *Pentre-mites*, with its distinctive five grooves from the mouth opening down the side of the body. With such a shape, it is not surprising that they are often mistaken for fossil hickory nuts. Like crinoids, they formed marine associations with corals, bryozoans, and brachiopods. Blastoids disappeared before the end of the Paleozoic era along with the never very successful cystoids and nearly all crinoid species.

Fossil crinoids are distinguished from blastoids and cystoids by the more substantial nature of crinoid stems and by possession of elaborately feathered arms. The body of a typical Paleozoic crinoid is covered by a symmetrical arrangement of limy plates. This body, the calyx, is topped by a mouth and a fringe of branched feathery arms. They resemble a stemmed flower complete with petals; hence they came to be called sea lilies. The feathery arms waved rhythmically in the water, supporting the animal and helping it to move. Crinoids, although they were sensitive to touch, lacked eyes. During the hundreds of millions of years that they flourished, they developed a great variety of size and shape of arm struc-

Stem segments of Pennsylvanian crinoids cover the surface of this hillside near Holdenville, Oklahoma.

ture and of the cross section of the stem, but by the end of the Paleozoic era they were almost extinct.

Classification of the more than 700 genera of fossil crinoids raises difficult problems about their evolutionary relationships. Classification is based on the structure of the plates of the calyx where they join the stem, on the branching of the arms, and the cross section and shape of the stem. The names of most genera are easily recognized because they end in the syllables—*crinus* or *crinites*.

The holothuroids (sea cucumbers) are echinoderms, but they do not play an important role as fossils because they have no readily preserved hard parts, other than microscopic spines embedded in their skin. They may be thought of as crinoids that have fallen on their sides. Starfish, which first appeared in the Ordovician rocks, break up easily, like crinoids, and for this reason well-preserved specimens are moderately rare. They have been likened to crinoids that abandoned the stalked way of life, turned over on their faces, and developed feet.

Primitive starfish had broad arms covered with feathery structures like the arms of crinoids. In later types the arms became more slender and the animal developed tube feet. The ophiuroids, or brittle stars, changed the pattern further with a body disc from which the five slender whiplike arms writhed. With these arms the starfish pulls itself along and grasps its prey.

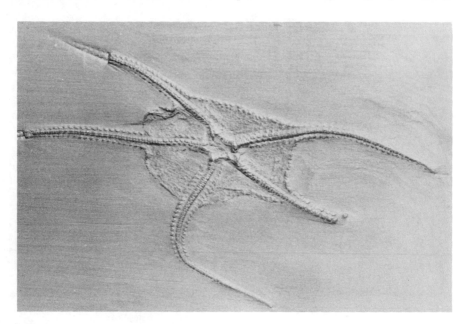

One of the many fine fossils of Devonian starfish in slate from Bundenbach, Germany.

Echinoids (sea urchins) and their flattened kin, the sand dollars, build hollow, boxlike skeletons of limy plates to which rods or spines are attached with articulated joints. Five food grooves radiate from the center of the upper side toward the mouth on the lower side. The spines protect the animal, gather food, and are the organs for stilted walking. The points of attachment are marked with small knobs on the surface of the skeleton. For grinding food, the urchins evolved a curiously complicated toothed mechanism known as "Aristotle's lantern," which works somewhat like a garbage-disposal unit.

In the Jurassic period, echinoids displayed rebellious evolutionary traits that produced such irregular forms as the furry sand dollars and the burrowing heart urchins of today. Like some other fossil echinoderms, early sea urchins are not well preserved, but later forms had tougher frameworks that survived, becoming common fossils in Cretaceous rocks.

Graptolites

The graptolites, constituting a mysterious and extinct phylum of sea animals, occupy the apex of the invertebrate world. Their fossils are little more than a carbon film in black slates and shales of the early Paleozoic era. They got their name, which means "writing on stone," because they resemble pencil marks. Detailed investigation revealed that these tracings had been complex colonies of small cuplike animals with a hollow nerve cord like a rudimentary spinal column. The graptolite colonies, at least in some species, drifted just below the surface of the sea buoyed up by a bladderlike float.

Their visceral resemblance to living marine wormlike animals known as pterobranchs, and to sea squirts and acorn worms, which may be in the line of ancestry of the vertebrates, gives them a much greater significance than their insignificance as fossils had suggested. The graptolites are now classified as fossils akin to these living forms, even though there remains much to learn about them.

One graptolite genus known in North America is *Dictyonema*, from the Ordovician period, which was the zenith of graptolite development. It grew as a lacy structure of branches connected with crossbars into a bell-like shape. The bell probably hung by a thread from a disc. The branches were made up of short tubes covered with a chitinous material. Some other graptolites apparently had a disclike foot that held them to the sea bottom.

Graptolites, which disappeared in the Mississippian period, are found more commonly in black shale laid down where oxygen was lacking than in other rocks. They are the only common fossil of many hard black shales.

Not a fossil hacksaw blade, but the serrated fossil graptolite *Tetrograptus*. Most grapolite fossils are scarcely wider than a pencil line and are usually preserved as a black carbon film. This unusual specimen is white on black Silurian shale and is from Norway.

These are the major divisions of invertebrates, from the algae to the threshold of the vertebrate phylum that includes man. After untold and unknown millions of years of almost unrecorded development, nearly all the phyla suddenly made their marks in the rocks of the Cambrian period, filled the oceans with teeming life, and brought forth hardy species that ventured onto land and prepared the way for life as we know it today.

The diversity of animal life increased steadily from the Cambrian into the Silurian period, fluctuated into the Permian that ended the Paleozoic era, declined in the Triassic of the Mesozoic era, then expanded from the Jurassic period into the Cenozoic era. Today it is much greater than at any time before the end of the Mesozoic era. The great procession has been marching a long way and a long time, but it is still going strong.

APPENDIX

STATE MAPS SHOWING FOSSIL AREAS

    Maps on the following pages were drawn for this book by Betty Crawford of Mansfield, Ohio. Included are maps of all states in the continental United States which have extensive areas of fossil-bearing rocks. Most states in the Northeast have few or no fossiliferous rocks, hence are not included in the maps.

    These maps locate areas where the rocks exposed or near the surface are of such a nature that they may contain fossils of interest to the collector. Not only will these maps serve as a guide to such areas, but they are useful to approximate the geologic age of any fossil found in place. This can be determined by finding on the maps the locality where the fossil was collected, and then using the key below. The maps are necessarily general in their descriptions and should be supplemented with large-scale geologic maps for precise identification, particularly in mountainous regions where outcrops are often small. Once the approximate age of a fossil is determined in this way, identification is much easier.

    The maps appear in the alphabetical order of the names of the states. Some maps, however, include more than one state. Use the list below to find the maps you wish to consult.

ALABAMA

GEORGIA

Quaternary

Pliocene

Miocene

Oligocene

Eocene

Cretaceous

Pennsylvanian

Mississippian

Cambrian - Ordovician

No fossils

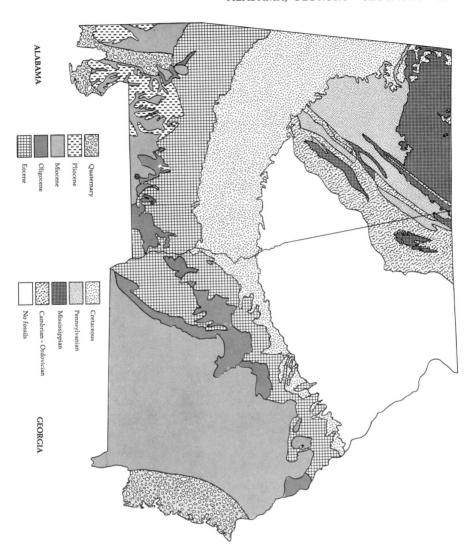

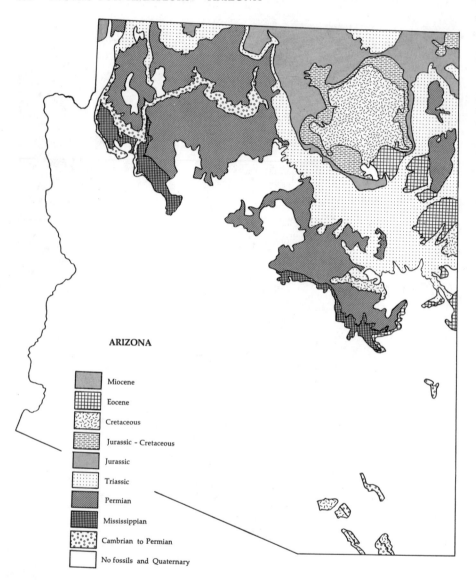

ARIZONA

- Miocene
- Eocene
- Cretaceous
- Jurassic - Cretaceous
- Jurassic
- Triassic
- Permian
- Mississippian
- Cambrian to Permian
- No fossils and Quaternary

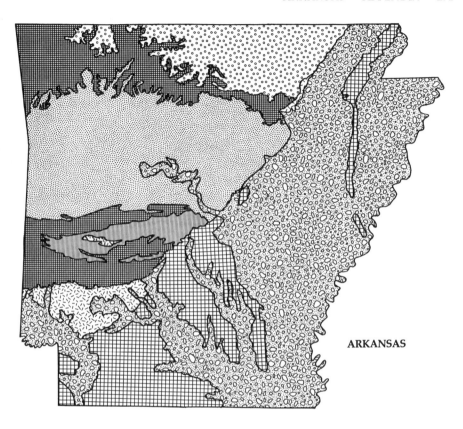

ARKANSAS

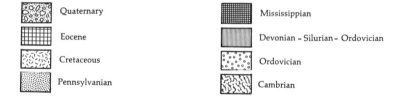

	Quaternary		Mississippian
	Eocene		Devonian - Silurian - Ordovician
	Cretaceous		Ordovician
	Pennsylvanian		Cambrian

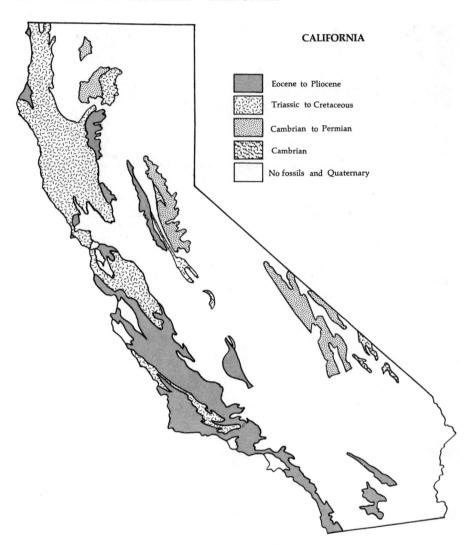

CALIFORNIA

Eocene to Pliocene

Triassic to Cretaceous

Cambrian to Permian

Cambrian

No fossils and Quaternary

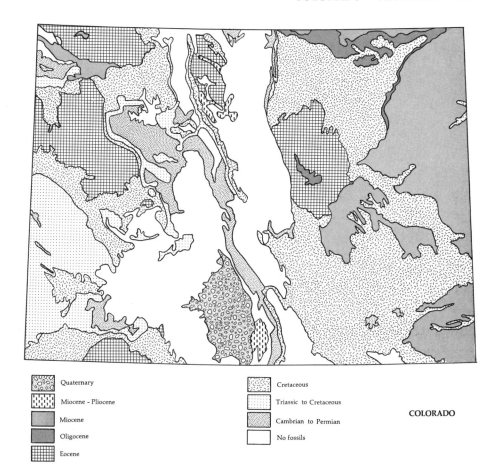

Quaternary

Miocene - Pliocene

Miocene

Oligocene

Eocene

Cretaceous

Triassic to Cretaceous

Cambrian to Permian

No fossils

COLORADO

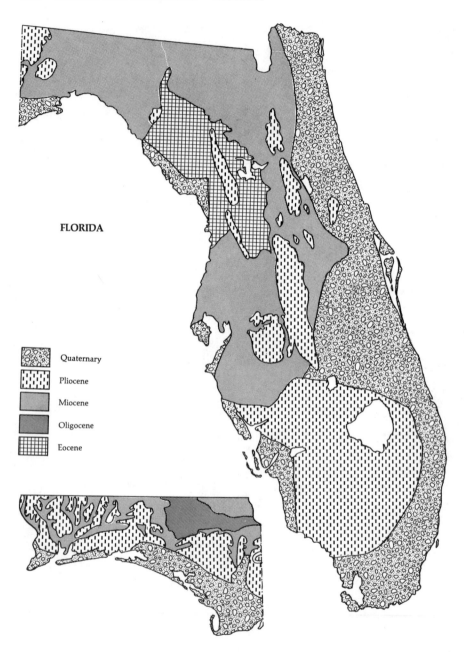

FLORIDA

Quaternary

Pliocene

Miocene

Oligocene

Eocene

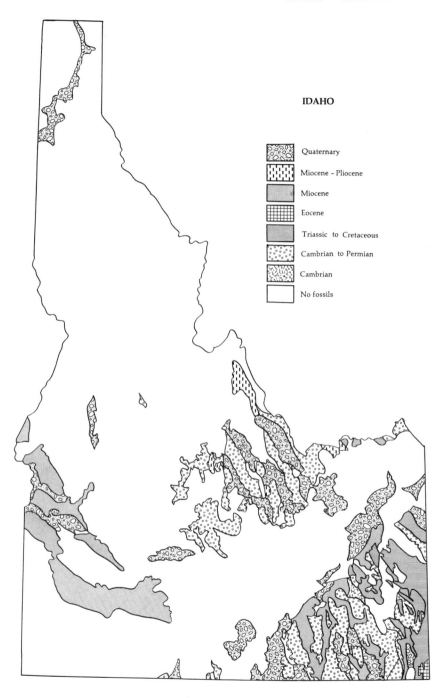

IDAHO

Quaternary

Miocene - Pliocene

Miocene

Eocene

Triassic to Cretaceous

Cambrian to Permian

Cambrian

No fossils

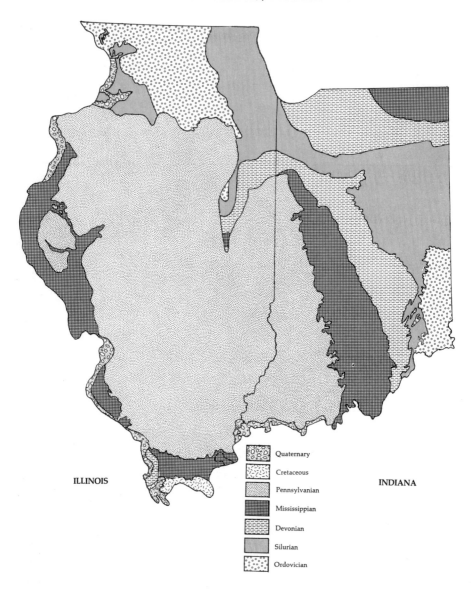

Quaternary
Cretaceous
Pennsylvanian
Mississippian
Devonian
Silurian
Ordovician

ILLINOIS

INDIANA

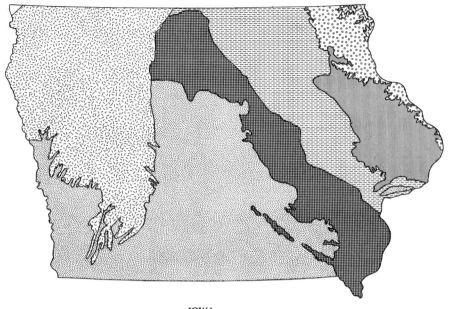

IOWA

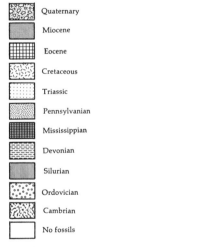

Quaternary

Miocene

Eocene

Cretaceous

Triassic

Pennsylvanian

Mississippian

Devonian

Silurian

Ordovician

Cambrian

No fossils

NEW JERSEY

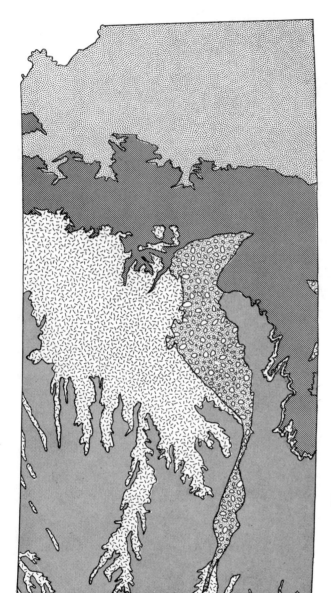

KANSAS

Quaternary

Miocene

Cretaceous

Permian

Pennsylvanian

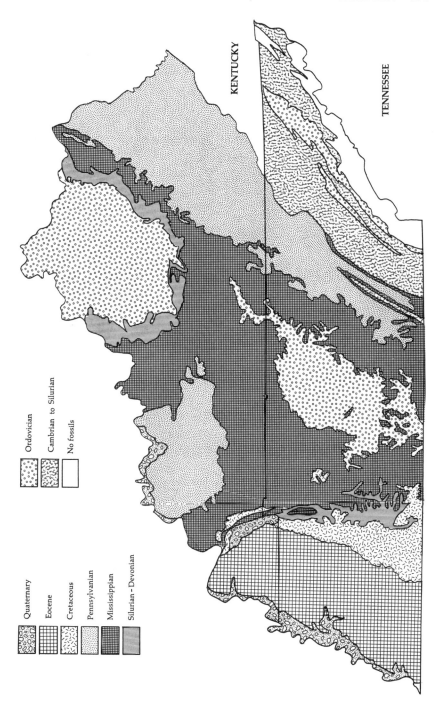

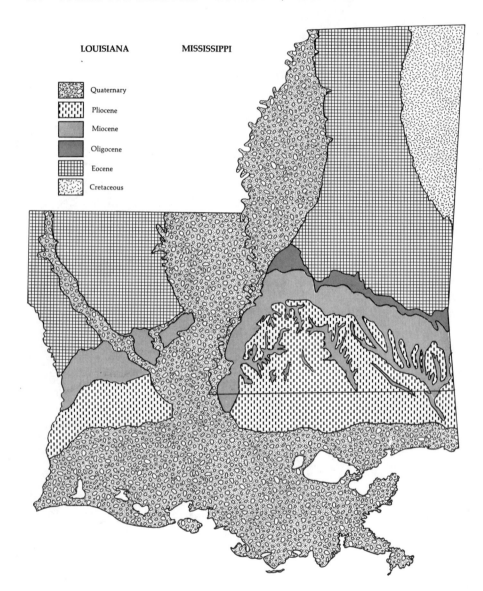

LOUISIANA

MISSISSIPPI

Quaternary

Pliocene

Miocene

Oligocene

Eocene

Cretaceous

MICHIGAN

Jurassic

Pennsylvanian

Mississippian

Devonian - Mississippian

Devonian

Silurian

Ordovician

Cambrian

No fossils

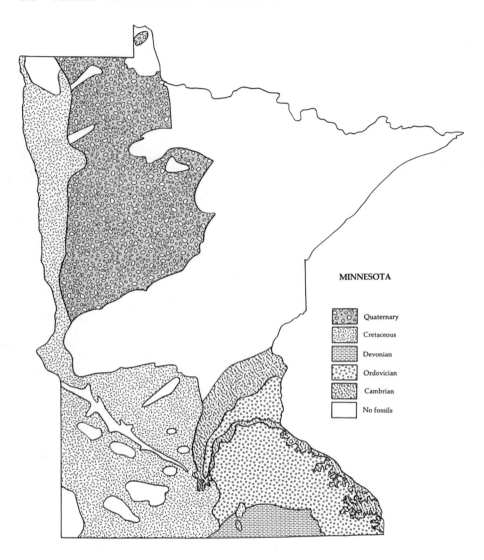

MINNESOTA

Quaternary

Cretaceous

Devonian

Ordovician

Cambrian

No fossils

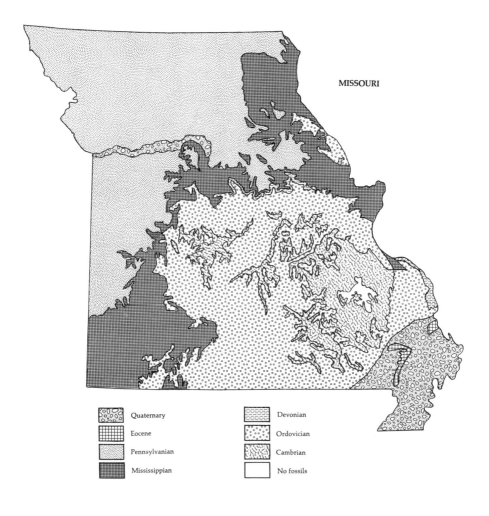

MISSOURI

Quaternary

Eocene

Pennsylvanian

Mississippian

Devonian

Ordovician

Cambrian

No fossils

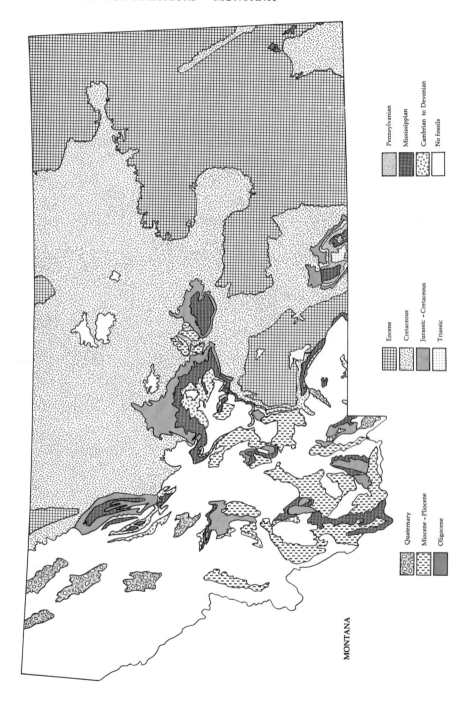

MONTANA

Eocene

Cretaceous

Jurassic – Cretaceous

Triassic

Quaternary

Miocene – Pliocene

Oligocene

Pennsylvanian

Mississippian

Cambrian to Devonian

No fossils

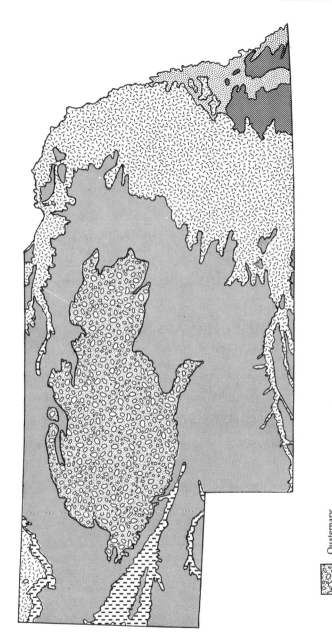

NEBRASKA

Quaternary

Miocene

Oligocene

Cretaceous

Permian

Pennsylvanian

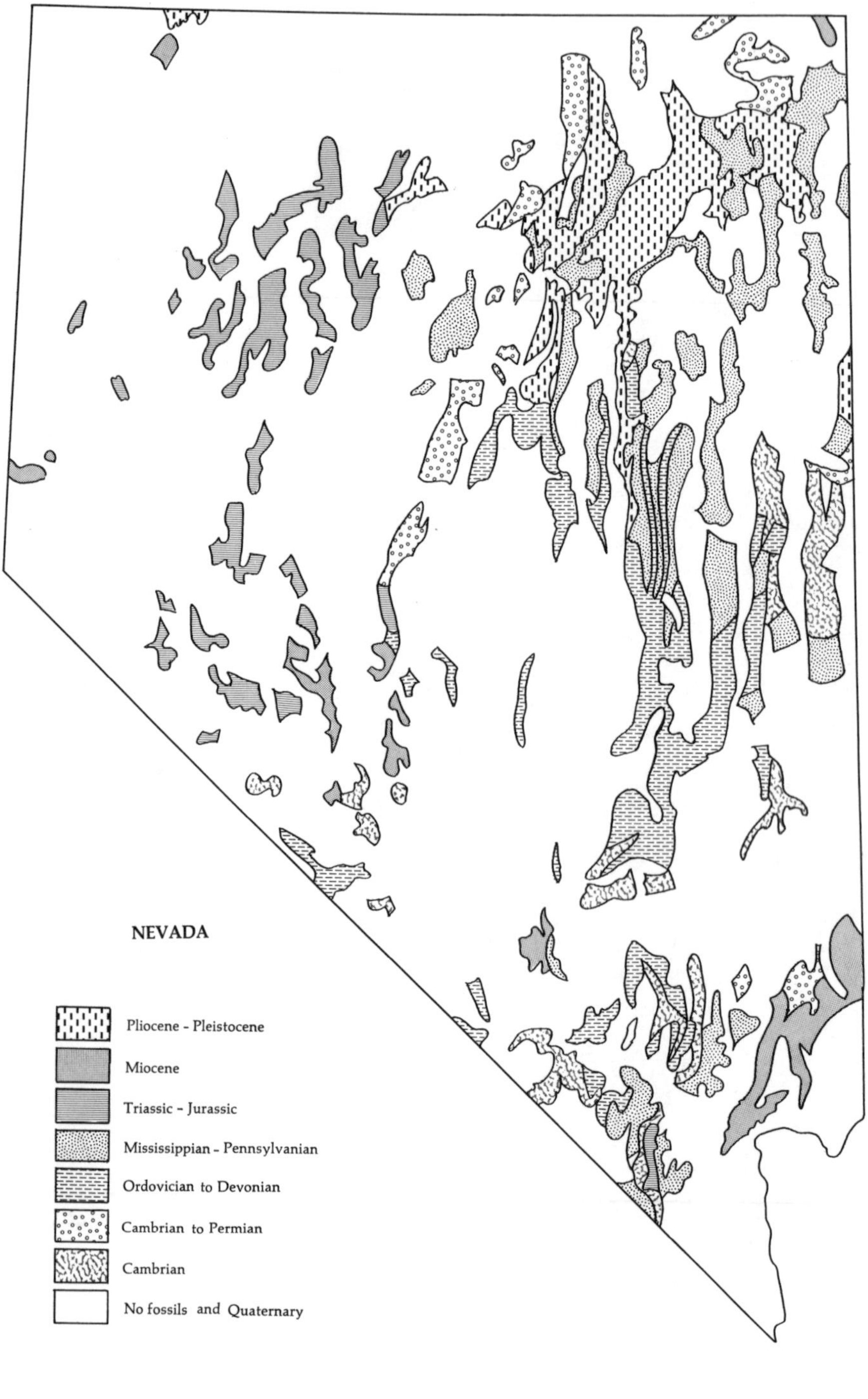

NEVADA

Pliocene - Pleistocene

Miocene

Triassic - Jurassic

Mississippian - Pennsylvanian

Ordovician to Devonian

Cambrian to Permian

Cambrian

No fossils and Quaternary

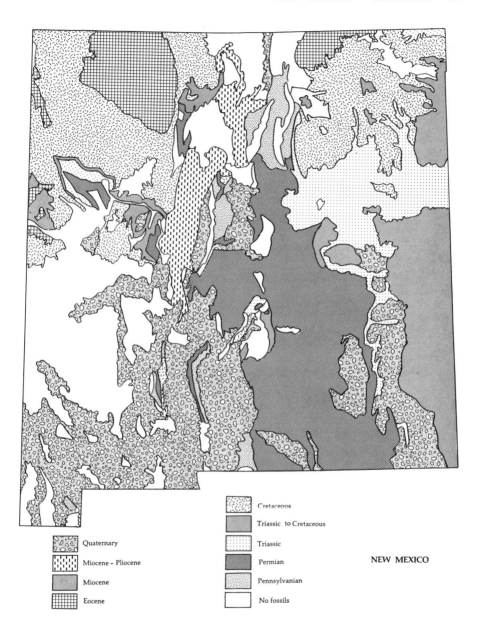

Cretaceous

Triassic to Cretaceous

Quaternary

Triassic

Miocene - Pliocene

Permian   **NEW MEXICO**

Miocene

Pennsylvanian

Eocene

No fossils

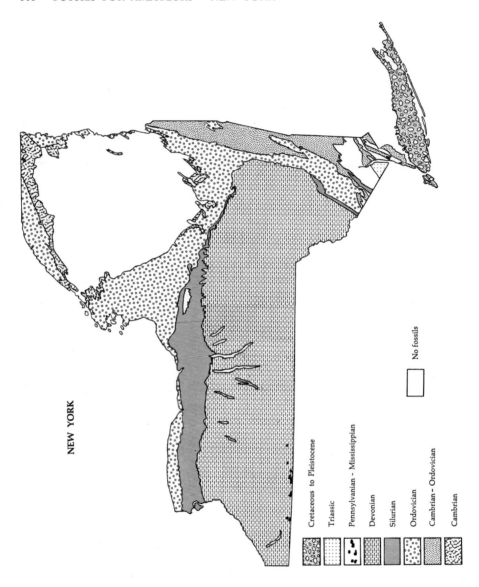

NEW YORK

Cretaceous to Pleistocene

Triassic

Pennsylvanian - Mississippian

Devonian

Silurian

Ordovician

Cambrian - Ordovician

Cambrian

No fossils

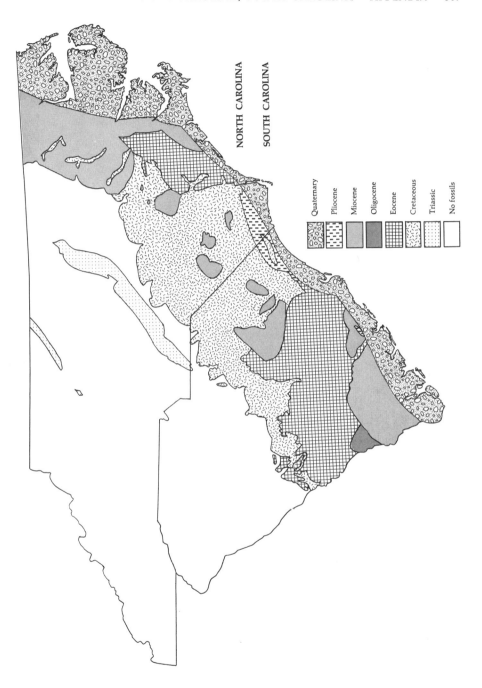

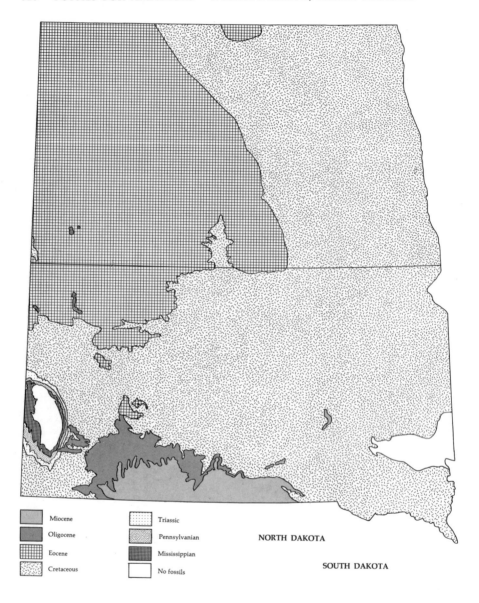

	Miocene		Triassic
	Oligocene		Pennsylvanian
	Eocene		Mississippian
	Cretaceous		No fossils

NORTH DAKOTA

SOUTH DAKOTA

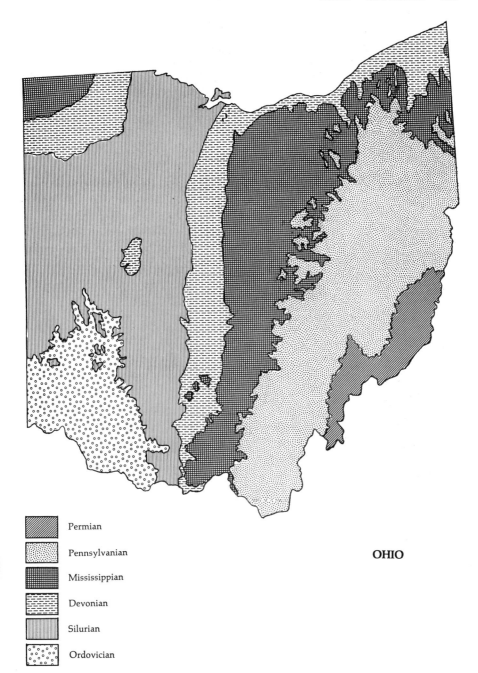

Permian

Pennsylvanian

Mississippian

Devonian

Silurian

Ordovician

OHIO

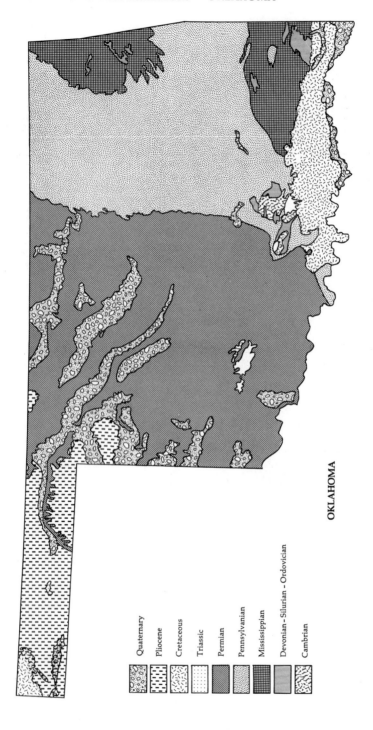

OKLAHOMA

Quaternary

Pliocene

Cretaceous

Triassic

Permian

Pennsylvanian

Mississippian

Devonian - Silurian - Ordovician

Cambrian

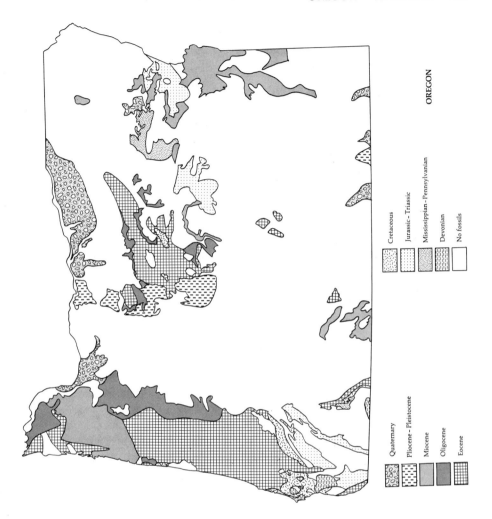

OREGON

Cretaceous

Jurassic – Triassic

Mississippian – Pennsylvanian

Devonian

No fossils

Quaternary

Pliocene – Pleistocene

Miocene

Oligocene

Eocene

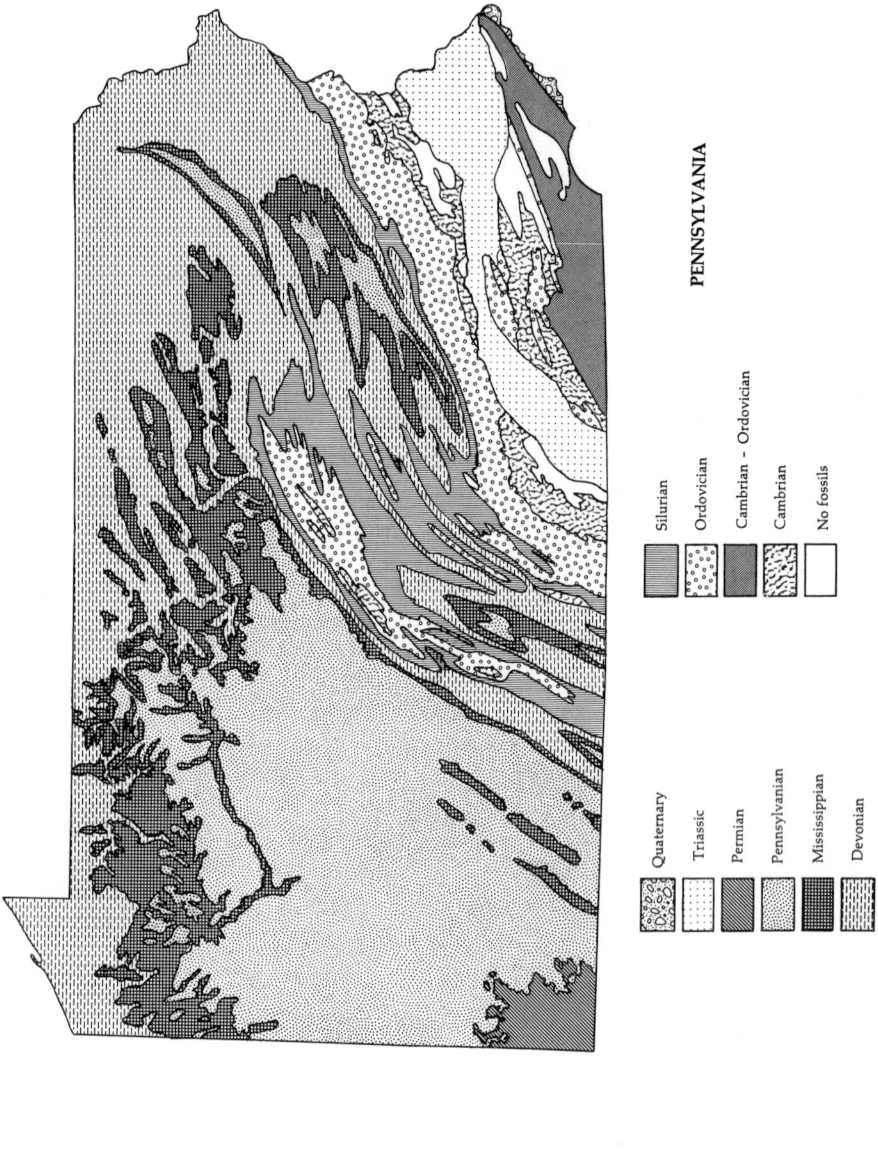

PENNSYLVANIA

Quaternary

Triassic

Permian

Pennsylvanian

Mississippian

Devonian

Silurian

Ordovician

Cambrian - Ordovician

Cambrian

No fossils

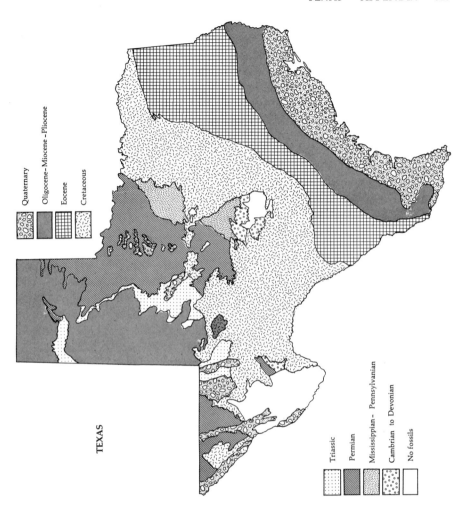

TEXAS

Quaternary

Oligocene – Miocene – Pliocene

Eocene

Cretaceous

Triassic

Permian

Mississippian – Pennsylvanian

Cambrian to Devonian

No fossils

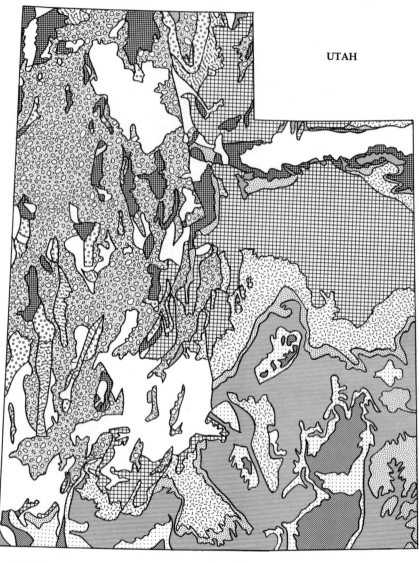

UTAH

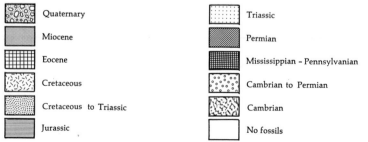

Quaternary		Triassic
Miocene		Permian
Eocene		Mississippian – Pennsylvanian
Cretaceous		Cambrian to Permian
Cretaceous to Triassic		Cambrian
Jurassic		No fossils

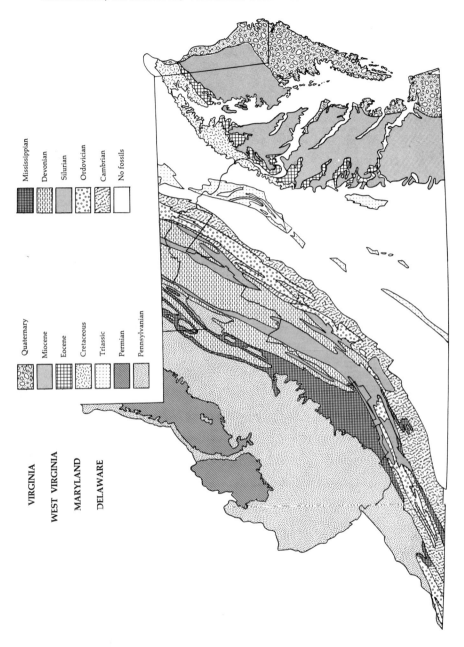

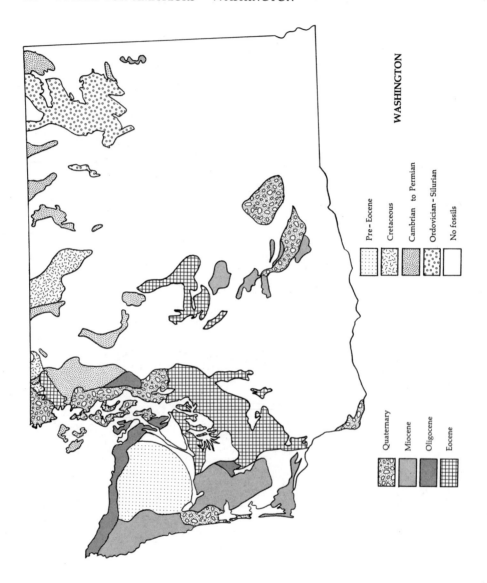

WASHINGTON

Pre - Eocene

Cretaceous

Cambrian to Permian

Ordovician - Silurian

No fossils

Quaternary

Miocene

Oligocene

Eocene

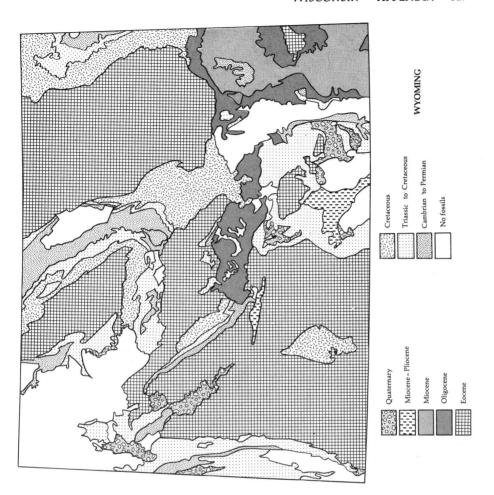

WYOMING

Quaternary
Miocene - Pliocene
Miocene
Oligocene
Eocene

Cretaceous
Triassic to Cretaceous
Cambrian to Permian
No fossils

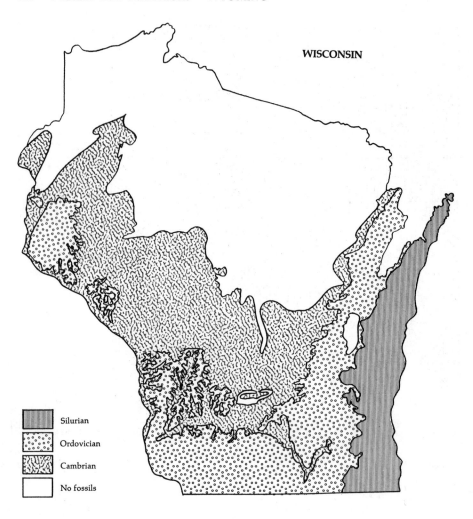

WISCONSIN

Silurian

Ordovician

Cambrian

No fossils

PUBLIC SOURCES OF FURTHER INFORMATION BY STATE

(See also "Specific Location Guides" under Recommended Books)

Alabama: Geological Survey, P.O. Drawer O, University 35486.

Alaska: Dept. of Natural Resources, Division of Mines and Geology, P.O. Box 5-300, College 99701.

Arizona: Bureau of Mines, University of Arizona, Tucson 85721.

Arkansas: Geological Commission, State Capitol, Little Rock 72119.

California: Division of Mines and Geology, Resources Building, Room 1341, 1416 Ninth Street, Sacramento 95814.

Colorado: Bureau of Mines, 316 State Services Building, Denver 80202.

Connecticut: Geological and Natural History Survey, Box 128, Wesleyan Station, Middletown 06457.

Delaware: Geological Commission, University of Delaware, Newark 19711.

Florida: Geological Survey, Box 631, Tallahassee 32304.

Georgia: Department of Mines, Mining and Geology, Agricultural Laboratory Building, 19 Hunter St., S.W., Atlanta 30334.

Hawaii: claims to have no fossils.

Idaho: Department of Public Lands, Boise 83707.

Illinois: Geological Survey, Natural Resources Building, Urbana 61801.

Indiana: Geological Survey, 611 N. Walnut Street, Bloomington 47401.

Iowa: Geological Survey, Geological Survey Building, Iowa City 52240.

Kansas: Geological Survey, University of Kansas, Lawrence 66044.

Kentucky: Department of Natural Resources, 209 St. Clair Street, Frank-fort 40601.

Louisiana: Geological Survey, Box 8847, University Station, Baton Rouge 70803.

Maine: Department of Economic Development, State House, Augusta 04330.

Maryland: Geological Survey, Latrobe Hall, Johns Hopkins University, Baltimore 21218.

Massachusetts: no geological agency.

Michigan: Department of Natural Resources, Stevens T. Mason building, Lansing 48926.

Minnesota: Geological Survey, University of Minnesota, Minneapolis 55414.

Mississippi: Geological Survey, 2525 North West Street, Jackson 39205.

Missouri: Division of Geological Survey and Water Resources, Box 250, Rolla 65401.

Montana: Bureau of Mines and Geology, College of Mineral Science and Technology, Butte 59701.

Nebraska: Conservation and Survey Division, University of Nebraska, 113 Nebraska Hall, Lincoln 68508.

Nevada: Bureau of Mines, University of Nevada, Reno 89507.

New Hampshire: Department of Resources and Economic Development, Division of Economic Development, James Hall, University of New Hampshire, Durham 03824.

New Jersey: Department of Conservation and Economic Development, Bureau of Geology and Topography, Box 1889, Trenton 08625.

New Mexico: Bureau of Mines and Mineral Resources, Campus Station, Socorro 87801.

New York: University of the State of New York, State Museum and Science Service, Albany 12224.

North Carolina: Mineral Resources Division, Department of Conservation and Development, Raleigh 22607.

North Dakota: State Geologist, University of North Dakota, Grand Forks 58202.

Ohio: Geological Survey, 155 South Oval Drive, Ohio State University, Columbus 43210.

Oklahoma: Geological Survey, University of Oklahoma, Norman 73069.

Oregon: Department of Geology and Mineral Industries, 1069 State Office Building, Portland 97201.

Pennsylvania: Bureau of Topographic and Geologic Survey, Main Capitol Annex, Harrisburg 17120.

Rhode Island: Department of Natural Resources, Veterans' Memorial Building, Providence 02903.

South Carolina: State Development Board, Division of Geology, Box 927, Columbia 29202.

South Dakota: Geological Survey, Science Center, University of South Dakota, Vermillion 57069 or Department of Highways, Pierre 57501.

Tennessee: Department of Conservation, Division of Geology, G-5 State Office Building, Nashville 37219.

Texas: Highway Department, 11th and Brazos Street, Austin 78711.

Utah: Geological and Mineralogical Survey, 200 Mines Building, University of Utah, Salt Lake City 84102.

Vermont: Geological Survey, East Hall, University of Vermont, Burlington 05401.

Virginia: State Geologist, Division of Mineral Resources, Box 3667, Charlottesville 22903.

Washington: Department of Natural Resources, Box 168, Olympia 98501.

West Virginia: Geological and Economic Survey, Box 879, Morgantown 26505.

Wisconsin: Geological and Natural Survey, 1815 University Avenue, Madison 53706.

Wyoming: Geological Survey, University of Wyoming, Box 3008, University Station, Laramie 82070.

DEALERS IN FOSSILS

Geological Enterprises, Box 996, Ardmore, Oklahoma 73401
Specialist in rare and unusual fossils from many localities. Has very wide variety of very fine specimens. Catalog $1.50; museum specimen catalog $1.00.

Fossils Unlimited, 9925 Highway 80 west, Fort Worth, Texas 76116
Good selection of fossils, primarily Texas Cretaceous. Free catalog.

Malicks Fossils, 5514 Plymouth Road, Baltimore, Maryland 21214
Very large variety of fossils. Supplier to many schools. Has over 6000 species described in catalog, $2.00.

Mid-America Rock Shop, 6928 North Clark Street, Chicago, Illinois 60626
Good variety of all types of fossils.

Glen's Gems, 139 North Lassen Street, Willows, California 95988
  Fossils of all types.
Ray's Fossils, Box 1364, Ocala, Florida 32670
  Wide variety of Florida fossils. Mail-order only.
Russel's Fossils, P.O. Box 71, Worcester, Massachusetts 01602
  Good variety of fossils from many localities.
Ward's Natural Science Establishment, P.O. Box 1712, Rochester, New
  York 14603
  Primarily a supplier of specimens and sets to schools. Has hundreds of
  species sold individually, as well as collections. Very large catalog; write
  for price.
In addition, many rock shops carry limited numbers of fossil specimens,
  often locally collected. A complete list of all North American rock shops
  is carried in each April issue of the *Lapidary Journal.*

RECOMMENDED BOOKS

For General Reading

Adams, Alexander B. *Eternal Quests.* New York: Putnam, 1968
  The story of the great naturalists. Well-written biographies, including
  many of the giants of paleontology.
Darwin, Charles. *On the Origin of the Species.* Various editions, 1859.
  A classic on evolution.
Eiseley. Loren C. *The Immense Journey.* New York: Random House, 1946.
  A work of literature on the emergence and progress of life.
Geikie, Sir Archibald. *The Founders of Geology.* New York: Dover paper-
  back, 1962 (first pub. 1897).
  A master of science and literary style tells of the men who discovered
  secrets of the earth.
Hapgood, Charles H. *Earth's Shifting Crust.* New York: Pantheon, 1958.
  A fascinating theory, carefully documented, about how the crust of the
  earth is able to slip great distances, causing ice ages and great extinctions
  of life.
Krutch, Joseph Wood. *Grand Canyon.* New York: Sloane, 1958.
  A noted writer describes the Grand Canyon and its geological sig-
  nificance.
Ley, Willy. *Dragons in Amber.* New York: Viking, 1951.
  Interesting and complete story of amber.
Miller, Hugh. *The Old Red Sandstone.* Various editions; first pub. 1841.
  Superb writing, masterpiece of the man who was perhaps geology's
  greatest writer.

Osborn, Henry F. *Men of the Old Stone Age*. New York: Scribner, 3d ed. 1919.
 Superb writing about ancient man and his animal neighbors.
Wendt, Herbert. *Before the Deluge*. New York: Doubleday, 1968.
 Popular account of geology and its founders, written in an anecdotal style.

General Geology and Paleontology

Augusta, Josef and Burian. *Prehistoric Animals*. London: 1956.
 Handsomely illustrated and imaginatively written for the young.
Bayly, Brian. *Introduction to Petrology*. Englewood Cliffs, N.J.: Prentice-Hall, 1968.
 Excellent chapters on sedimentary rocks.
Beerbower, James R. *Search for the Past*. Englewood Cliffs, N.J.: Prentice-Hall, 1960.
 One of the most readable, witty nontechnical books, well illustrated with drawings of both vertebrate and invertebrate fossils, posing many unsolved problems of paleontology.
Brouwer, A. *General Paleontology*. U. of Chicago Press, 1959.
 Excellent on paleontological theory, by a Dutch scholar.
Buchsbaum, Ralph. *Animals Without Backbones*. U. of Chicago Press, 2d ed. 1948.
 Highly readable text about living invertebrates, valuable in understanding relationships among fossil invertebrates.
Casanova, Richard. *An Illustrated Guide to Fossil Collecting*. Healdsburg, Calif.: Naturegraph, new ed. 1970.
 A good place to start, with much practical advice and a short list of locations.
Clark, T. H., and C. W. Stern. *Geological Evolution of North America*. New York: Ronald, 2d ed. 1960
 Good general textbook with much about fossils.
Colbert, Edwin H. *Dinosaurs*. New York: Dutton, 1961.
 Tops in its field.
————. *Men and Dinosaurs*. New York: Dutton, 1968.
 The great dinosaur hunters and their prey.
Croneis, Carey, and W. C. Krumbein. *Down to Earth*. U. of Chicago Press, 1936.
 An introductory textbook that is informal and yet covers the subject.
Dunbar, Carl O. *Historical Geology*. New York: Wiley, 2d ed. 1960.
 College-level textbook with emphasis on fossils. Strong on definition and the physical geology of paleontology.

Fenton, Carroll and Mildred. *The Fossil Book*. New York: Doubleday, 1958.
Superbly illustrated, an excellent nontechnical general discussion of fossils—what they are and how they occur. It has general value as an identification text.

Flint, Richard P. *Glacial and Pleistocene Geology*. New York: Wiley, 1957.
Good chapter on Pleistocene fossils and good general background.

Hurley, Patrick M. *How Old Is the Earth*. New York: Anchor paperback, 1959.
Readable, nontechnical discourse on geologic time and how it is determined.

King, Philip B. *The Evolution of North America*. Princeton U. Press, 1959.
Scholarly study of the geological history of the continent.

Kummel, Bernhard. *History of the Earth*. San Francisco: Freeman, 2d ed. 1970.
College-level textbook.

Matthews, W. H. III. *Wonders of Fossils*. New York: Dodd, 1968.
A good book for the beginning young collector.

———. *Fossils: An Introduction to Prehistoric Life*. New York: Barnes & Noble, paperback 1962.
Recommended introductory book on paleontology, acceptably illustrated and well rounded in material and presentation.

Moore, Raymond C. *Introduction to Historical Geology*. New York: McGraw-Hill, 2d ed. 1958.
One of the best college-level textbooks by an outstanding authority on fossils, well illustrated.

Moore, Ruth. *Time, Man and Fossils*. New York: Knopf, 1953.
Good on evolutionary theory and geological time dating.

Oakley, Kenneth P. *Frameworks for Dating Fossil Man*. Chicago: Aldine, 2d ed. 1966.
Comprehensive study of the dating of sedimentary rock deposits.

Richards, Horace. *Record of the Rocks*. New York: Ronald, 1953.
Excellent on the geology of the eastern United States; well illustrated.

Scheele, William E. *Prehistoric Animals*. New York: World, 1954.
Handsomely illustrated account of all ancient animals except mammals. For young people.

Simpson, George Gaylord. *Life of the Past*. Yale U. Press, paperback, 1953.
Perhaps the most enjoyable introduction to paleontology. An outstanding scientist explains his subject for the general reader and student.

Stirton, R. A. *Time, Life and Man*. New York: Wiley, 1959.
Readable college text also of interest to adult readers.

Fossil Identification and Specialized Resource Books

Andrews, H. N. Jr. *Studies in Paleobotany*. New York: Wiley, 1961.
Reference work on fossil plants.

Arnold, C. A. *Introduction to Paleobotany*. New York: McGraw-Hill, 1947.
Nontechnical guide to a difficult field; well illustrated.

Camp, C. L., and G. D. Hanna. *Methods in Paleontology*. U. of Calif. Press, 1937
Helpful for collecting and preparing fossils.

Case, Gerard. *Fossil Shark and Fish Remains of North America*. Self-published, 1969.
Well-illustrated identification guide to fossil fish likely to be found by the amateur collector.

Cushman, Joseph. *Foraminifera*. Harvard U. Press, 4th ed. 1948.
Reference work of value in understanding these microfossils.

Darrah, W. C. *Principles of Paleobotany*. New York: Ronald, 2d ed. 1960.
College text.

Glaessner, M. F. *Principles of Micropaleontology*. New York: Wiley, 1947.
Advanced guide to techniques.

Jones, Darnell J. *Introduction to Microfossils*. New York: Harper, 1956.
Useful for techniques of collection and identification of the minifossils.

Knowlton, Frank H. *Plants of the Past*. Princeton U. Press, 1957.
Nontechnical account of fossil plants.

Kummel, Bernhard, and David Raup. *Handbook of Paleontological Techniques*. San Francisco: Freeman, 1955.
Advanced text of some general use, primarily designed for museums and universities with extensive equipment.

Langford, George. *Wilmington Coal Flora*. ESCONI Associates, Chicago, 1958.
The most extensive, well-illustrated identification book for Pennsylvanian plant fossils found in the Midwest.

Moore, Raymond C., C. G. Lalicker, and A. G. Fisher. *Invertebrate Fossils*. New York: McGraw-Hill, 1953.
College text, many useful illustrations of common fossils

Moore, Raymond C. *Treatise of Invertebrate Paleontology*. U. of Kansas and Geol. Society of America, various dates.
A lengthy series of books, highly technical, each treating a major group of invertebrate fossils. They are designed to illustrate and discuss every genus of invertebrate fossils known in the world. The series is almost complete, and is the ultimate general series for fossil determination to generic level.

Romer, Alfred. *Man and the Vertebrates*. Baltimore: Penguin paperback, 2 vols., 1954 (1st pub. 1933).

Excellent account of man's place in the scheme of things.

———— *Vertebrate Paleontology*. U. of Chicago Press, 3rd ed. 1966.

The classic college text on fossils with backbones, useful also for identification of some vertebrate fossils with its excellent illustrations.

Shimer, H. W., and R. R. Shrock. *Index Fossils of North America*. New York: Wiley, 1944.

The major single-book source for identification of fossils to specific level. Probably 90 percent of the invertebrate fossils found by the average collector can be identified using this book, which has over 700 pages of illustrated fossils.

Shrock, R. R., and W. H. Twenhofel. *Principles of Invertebrate Paleontology*. New York: McGraw-Hill, 1953.

Advanced textbook, well illustrated, of value in identification.

Thomas, R. C. *Let's Find Fossils on the Beach*. Pub. by author, Venice, Florida, 1961.

A very helpful illustrated guide for identification of the teeth, bones, and other vertebrate remains found on the beaches of the East Coast from New Jersey to Florida.

Wood, Henry. *Invertebrate Paleontology*. Cambridge U. Press, 1961 (first pub. 1897).

Inexpensive paperback reference book, by an English authority.

General Location Guides

Murray, Marian. *Hunting for Fossils*. New York: Macmillan, 1967.

A state-by-state account of specific fossil sites, with a good bibliography of state publications.

Ransom, J. E. *Fossils in America*. New York: Harper, 1964.

A state-by-state compilation of thousands of locations gleaned from other publications, some good, some bad.

Specific Location Guides

Alabama: *Curious Creatures in Alabama Rocks*, a guidebook by C. H. Copeland Jr. Geol. Surv. Ala. Circ. 19, 1963.

Arizona: *Paleontological Literature of Arizona*, by H. W. Miller Jr. U. of Ariz. Press, 1960.

Arkansas: *Fossils of Arkansas*. Bulletin 22, Arkansas State Geological Commission.

California: *Fossils, What they Mean and How to Collect Them.* Vol. 13, No. 7, Mineral Information Service, Division of Mines, San Francisco.
"A Plea for Fossil Vertebrates," by J. R. MacDonald. Vol. 17, No. 12, Mineral Information Service.
Rancho La Brea, by Chester Stock. Los Angeles County Museum, 4th ed. 1949.
Geologic Guidebook of San Francisco Bay Counties, by O. P. Jenkins. Division of Mines. Many fossil localities.
Colorado: *In My Back Yard,* by Al Look. U. of Denver Press, 1951. The dinosaur monument area by an amateur collector.
Fossils, by H. C. Markham. Denver Museum of Natural History popular series no. 3.
Connecticut: *Triassic Life of the Connecticut Valley,* by R. S. Lull. Geological and Natural History Survey Bulletin 81, 1953.
Florida: *Vertebrate Fossil Localities in Florida,* by S. J. Olson. Florida Geol. Survey Spec. Pub. No. 12, 1965.
Pliocene Fossils. Geol. Bulletin 40.
Illinois: *Field Book, Pennsylvanian Plant Fossils of Illinois,* Illinois State Geol. Survey Educ. Series No. 6.
Guide for Beginning Fossil Hunters. Ill. State Geol. Survey Educ. Series No. 4.
The Past Speaks to You, by Ann Livesay. State Museum, Story of Illinois No. 1, 1951.
Indiana: *Adventures with Fossils,* by Robert H. Shaver. Indiana Geol. Survey Circular No. 6, 1959.
Fossil Plants of Indiana, by J. E. Canright. Indiana Geol. Survey Reports of Progress No. 14, 1959.
Fossils: Prehistoric Animals in Hoosier Rocks, by T. G. Perry. Indiana Geol. Survey No. 7, 1959.
Iowa: *Fossils and Rocks of Eastern Iowa,* by J. N. Rose. Iowa Geol. Survey Educ. Series 1, 1967.
Maryland: *Miocene Fossils of Maryland,* by H. E. Vokes. Dept. of Geology Bulletin 20, 1957.
Michigan: *Guide to Michigan Fossils,* by R. W. Kelley. Dept. of Conservation.
Minnesota: *Minnesota's Rocks and Waters,* by G. Schwartz and G. Thiel. Minn. Geol. Soc. Bulletin No. 37, 1954.
Guide to Fossil Collecting in Minnesota. Geol. Survey Educ. Series No. 1.
Missouri: *Common Fossils of Missouri,* by A. G. Unklesbay. U. of Missouri Handbook No. 4, 1955.
Montana: *Rocks and Fossils of Glacier National Park.* U. S. Geol. Surv. Prof. Paper 294-K, 1959.

Nebraska: *Record in Rock,* a handbook of the invertebrate fossils of Nebraska, by Roger Pabian. Educ. Circular No. 1, U. of Nebr., 1970.

New Jersey: *Cretaceous Fossils of New Jersey,* by Horace G. Richards. Bureau of Geol. and Topography Bulletin 61, 1962.

New York: *Handbook of Paleontology for Beginners and Amateurs,* by Winifred Goldring. State Museum Handbook No. 9, 1929. Reprinted by Paleontological Research Labs, 109 Dearborn Pl., Ithaca, N.Y.
Popular Guide to the Nature and Environment of the Fossil Vertebrates of New York, by R. L. Moodie. State Museum Handbook No. 12, 1933.

Ohio: *Ohio Fossils,* by A. LaRocque and M. F. Marple. Ohio Geol. Bulletin 54, 1955.
Handbook for Teachers of Earth Science, Info. Circular No. 15.
Elementary Guide to the Fossils and Strata of the Ordovician in the Vicinity of Cincinnati, by Kenneth Caster. Cinc. Museum of Nat. Hist., 1955.

Pennsylvania: *Fossil Collecting in Pennsylvania,* by D. M. Hoskins. Topographic and Geol. Survey, General Report G. 40, 1964

South Dakota: *Midwest Gem Trails,* by June Zeitner. Mentone, Calif., Gem Books.

Texas: *Texas Fossils,* by William Matthews III. Bureau of Economic Geol., U. of Tex., 1960.

Washington: *Fossils in Washington,* by V. E. Livingston Jr. Dept. of Conserv. Information Circular No. 33, 1959.

Wisconsin: *Fossil Collecting in Wisconsin,* by M. E. Ostrow. Geol. Surv. 1961.
Silurian Trilobites of S.E. Wisconsin, by J. E. Emielity. Milwaukee Pub. Museum, 1963.

NOTE: The preceding list deals with general guides to fossil sites. Many more specific sites can be obtained from guidebooks for specific field trips, roadside geology booklets, and descriptions of specific formations and groups published by the U.S. and state geological surveys or the equivalent state agency. A listing of their current publications can be obtained by writing the agencies listed elsewhere in the Appendix.

Magazine Sources

Earth Science. Bimonthly, carrying many excellent articles on fossils. Box 550, Downers Grove, Ill. 60515. In addition, excellent booklets have been made from reprinted fossil articles, all illustrated and many containing fossil locations: *Fossils from Our Earth,* and *Fossils of the Mid-Continent,* each $1.00.

Journal of Paleontology. Published bimonthly by the Society of Economic Paleontologists, P.O. Box 979, Tulsa, Okla. 74101. Very technical, primarily describing new fossil species or describing fossils from a specific locality. Contains much good information on fossil sites and fossil identification.

Lapidary Journal. P.O. Box 2639, San Diego, Calif. 92112. April issue carries a complete listing of all rock shops (including fossils and equipment) in the USA, plus listing of all rock clubs in the world.

Miscellaneous

Bibliography of Earth Science Materials, Midwest Fed. of Mineralogical and Geological Societies, 1969. Comprehensive, detailed list of publications and maps dealing with fossils and minerals of the Midwest. Periodically revised.

Bibliography of North American Geology, available at most large libraries or universities. Contains listings of all articles dealing with geology, by author, state, and title. Published yearly. Very handy for finding localities.

Geological Society of America, Works in Print, booklet issued describing available material published by the GSA. P.O. Box 1719, Boulder, Colo. 80302.

Geology and Earth Sciences Sourcebook, American Geological Institute. New York: Holt, 1962. A resource book for teachers in secondary schools of earth science.

Mineral, Fossil and Rock Exhibits and Where to See Them, pamphlet published in 1960 by the American Geological Institute, 2101 Constitution Ave. N.W., Washington, D.C. Annotated listing of 12 Canadian and 167 American museums.

INDEX

absolute time, 74–78; compared to relative time, 69
acetic acid, glacial, 170–3, 190, 226
acetone: for mixture to harden shales, 146; in cleaning fossils, 176, 226, 227; in making peels, 184, 185; as solvent for liquid resins, 191
acids: cleaning fossils with, 150, 169–78, 184, 190, 225, 226–7; rule for mixing, 173, 178, 184; in making peels, 180, 181, 184
agatized fossils, 106, 112
age determination by nuclear methods, *see* atomic clock
Agricola, Georgius, 13
Alabama: state laws governing collecting, 129; geological map, 289; sources of further information, 321
Alaska: state laws governing collecting, 129; sources of further information, 321
alcohol, cleaning fossils with, 150, 176, 226, 227
algae agate, 233
algae fossils, 54, 72, 233, 235–6, 246, 250
amateur fossil collectors, value of, to science, 2, 3, 5–9

amber, 8, 31, 44, 279–80
American Association of Petroleum Geologists, Bulletin of, 83
American Museum of Natural History (New York City), 17
ammonites, 85, 105, 108, 112, 113, 148; in giant shale concretions, 65; in ordinary concretions, 106; making casts of, 195; giant Cretaceous, 268
ammonoids, 267
ammonium chloride for whitening fossils, 213
amphibians, 114–15
angiosperms, 242, 243–6
anhydrite fossils, 174
animal burrows, finding fossils in, 114
animals, classifications of, 246–86 *passim*
Anthracomedusa turnbulli ("Coal Age jellyfish of Turnbull"), 5
Antiquities Act of 1906, 127
ants, primitive, 8
Archaeopteryx, 32
Archeozoic ("beginning life") era, 71, 78
Archimedes screws, 102, 255
Aristotle, 12